Niflaos HaBoreh
EXPLORER

ANATOMY AND PHYSIOLOGY

By Rabbi Yaakov Lubin, BSC

Editors:
Mrs. Sara Miriam Gross
Mrs. Chaya Silverstone

Graphic Design:
DesignByAviva@gmail.com

The Jewish Center for Science
3821 N 51st Street
Hollywood, Fl 33021
Phone: 1-954-416-3305
Email: YaakovLubin@jewishcenterforscience.com

ISBN-13: 978-1-7330531-1-2

This textbook is dedicated for the Refuah Shelema of

Lily Leah Madeb bat Virgine

by her husband Dr. Isaac Madeb

and their children:

Dr. Ralph and Limor Madeb

Dr. David and Marilyn Zarabi

Mr. Morris and Virna Zeitoune

Dr. Joseph and Sharon Safdieh

This textbook is dedicated by

Dr. Isaac and Lily Madeb

in memory of their parents:

Raffoul Mouhaddeb ben Garaz

Marie Mouhaddeb bat Jamilé

Moise Mann ben Rahel

Virginie bat Leila

This textbook is dedicated by

Andrew and Samantha Sossin

Their unwavering friendship
and support has enabled the dream
of making Torah-based textbooks
to come to fruition.

***May their sponsorship of this project
be a source of merit to bring them
and their children
health, happiness, and prosperity.***

This textbook is dedicated by

Mr. Leon Benlolo

in memory of and לע"נ

Yosef Tapiero OB"M

Freha bat Esther OB"M

Esther bat Freha OB"M

Abe and Lillian Tarler OB"M

Nathan A. Tarler OB"M

Mr. Benlolo's support of this project has been instrumental in bringing this textbook to fruition. May it be a source of great merit for him and his family and bring them health, happiness, and prosperity.

Sponsors

Yeshivat Mekor Haim

The Goldberg Family Foundation

Mr. Yaakov Mishan

Mr. Ari Pearl and Mr. Mickey Taillard

Patrons

Mr. Amram Adar

Mr. Ronnie Adjmi

Mr. Mayer Chemtob

Mr. Sam Chavin

Dr. Jacques Doueck

Mr. Freddie Erani and Mr. Shaul Doueck

Mr. Sydney Saadeh

Mr. Yehuda Seinfeld

Mr. Joseph Setton

Friends

Mr. Joey Levy

Mr. and Mrs. Boruch and Annette Goldberg

Mr. Zack Gindy

Mr. Max Mizrachi

Mr. Mark Shamma

The Dabah and Silverman family

TU
everyschool.
everyteacher.
everychild.
TORAHUMESORAH

Torah Umesorah Publications
620 Foster Avenue, Brooklyn, New York, 11230
Tel. 212-227-1000

28 Adar, 5778

I have reviewed portions of the book ***Fundamentals of Life Science,*** which is part of the "Niflaos HaBoreh Explorer" project of Rabbi Yaakov Lubin.

I am enthusiastic about the role that this volume can hopefully play in the teaching of science in Torah schools. It is no secret that the subject of science, in general, in the way it is portrayed within the general schoolbook industry, is problematic in many ways. In recent years, Torah schools have become increasingly sensitized to the challenges associated with using books from the secular world.

Works such as Rabbi Lubin's offer a wonderful alternative to those books, without sacrificing in the least the quality of the material, the scope of the coverage and the integrity of the study. It also has the added advantage of strengthening *emunah*.

I am excited to offer my endorsement to ***Fundamentals of Life Science,*** and I look forward to seeing future volumes by Rabbi Lubin aimed at providing materials that are written *al taharas hakodesh* and that meet the growing need in our *mosdos.*

With best wishes for *Hatzlacha...*

Rabbi Shmuel Yaakov Klein

Director of Publications

הרב אהרן פלדמן

RABBI AHARON FELDMAN
421 YESHIVA LANE, APT. 3A, BALTIMORE, MD 21208
TEL.: 410-6539433 FAX: 410-6534694
STUDY: 410-4847200 EXT. 6050; DIRECT LINE: 443-5486050
E-MAIL: RAF@NIRC.EDU

ROSH HAYESHIVA
NER ISRAEL RABBINICAL COLLEGE

ראש הישיבה
ישיבת נר ישראל

בס"ד

יט בתמוז תשע"ח/July 4, 2018

TO WHOM IT MAY CONCERN:

The Torah day school movement in the U.S. has been successful to a degree unimaginable 50 years ago: there are nearly 1000 such schools across the country. But what has been neglected in this school system is their general studies departments. Instead of using them as vehicles for inculcating Torah values and beliefs into our students, we permit them to transmit cultural norms and beliefs which are the antithesis of Torah.

One such area is the science departments of our Torah schools. Instead of using the wonders of the physical world as a means of teaching students that there is a Creator, we use textbooks (and often teachers) that accept as their premise the mathematical absurdity that existence came into being by itself by chance. We have been inexcusably negligent in failing to provide our students with textbooks which will inculcate belief in a Creator.

Rabbi Yaakov Lubin deserves the gratitude of the Torah community for having risen to the challenge. He has produced a science workbook, *The Fundamentals of Life Science,* for middle school level which is at once attractive, authoritative and yet conveys the message that behind all the wonders of science is a Creator Whose wisdom brought them into being.

I fervently hope that this science book will be embraced by schools across the land and that Rabbi Lubin, and others like him, will continue to produce such workbooks for all levels and subjects of Torah education.

With respect,
Aharon Feldman

Rabbi Aharon Feldman

RABBI MOSHE WOLFSON
1574 43RD STREET
BROOKLYN, NEW YORK 11219

בס"ד

מכתב המלצה

הנה מודעת זאת מפי ספרים ומפי סופרים ורבותינו מצוקי ארץ הגידו לנו מראש את אשר יקרה באחרית הימים. כאשר החושך יכסה ארץ, ושלהבת אור האמונה המושרשת בלב כל זרע אברהם אשר בשם ישראל יכונה- שבשנים כתיקונם היתה עולה מאליה ועוברת בירושה מדור לדור בכל תוקף זיוה והדרה- בעקבתא דמשיחא תהא נעדרת וטמונה בין הרי החושך, ורק ע"י יגיעה ועבודה מרובה ניתן לדלותה ולגלותה ממסתריה ממעמקי הלב אשר שם יתלחשון גחליה.

ועל כן בואו ונחזיק טיבותא לידידי וו"ח יראת ה' היא אוצרו הרה"ח ר' יעקב לובין שליט"א אשר רחש לבו הטהור ד'בר טוב לחזק ברכים כושלות, ולמטרה זו יסד מכון "ברכי נפשי" למען הקנות לצעירי הצאן את ערכי אמונה הטהורה ע"י הרחבת והמחשת בביאור מעשי ה' ונפלאותיו הסובבים אותנו תמיד, ובחכמת ה' וחסדיו המרובים אשר משפיע לבריאותיו בכל עת. וכידוע מרבותינו הרמב"ם ובעל חובת הלבבות ז"ל ועוד, כמה כוחו רב לגלות ולהבעיר את האמונה הגנוזה ולהביא לידי אהבת ה' ויראתו.

והנני מברכו שחפץ ה' בידו יצליח לגדל עדרים עדרים של צאן קדשים הרועים באמונה ומתעדנים מנאות דשאיה ומתיקות פירותיה, ובזכות האמונה נזכה במהרה לאור חדש על ציון ואז נשיר שיר חדש מראש אמנה על גאולתינו ועל פדות נפשינו

משה וואלפסאן

הרב שמעון גלאי
רב ומו"צ
רח' אהרונוביץ 51 בני ברק

בס"ד, תאריך: ז' אדר [illegible]

מכתב ברכה

הריני לחזק בזה את הרב יעקב לובין שליט"א שלקח
על עצמו להפיץ את המובא בחובת הלבבות
להתבונן בחסדי השי"ת שרואים מתוך הבריאה
ובאמת מהדברים שצריכים חיזוק גדול הוא
דבר זה, להתבונן בחסדיו בכל עת ובכל שעה
וכאמור זה נעשה ע"י חוברת אלבומי תמונות!
ומצוה גדולה לחזק אותם בכל מה שאפשר
שיצליחו להחדיר ענין זה בכל תפוצות ישראל
וכן באתי לחזק ידי הרב משה [illegible] שליט"א שמסייע
מאוד בענין זה

החותם בברכה,
שמעון גלאי

June 4, 2019

Letter of Approbation,

The spectacular textbooks entitled, *Niflaos Haboreh Explorer* by the esteemed, yirei shamaim, Rabbi Yaakov Lubin, are an unbelievable resource for the Elementary School Jewish student. These textbooks are the ultimate tool for teaching science in a Jewish environment, providing the highest level of kosher content, while maintaining an accurate and scientifically sound textbook.

The ability to breakdown scientific information and present it in a digestible & kosher manner, is not a simple task. Rabbi Lubin does a spectacular job making these esoteric and complicated scientific concepts, relatable and extremely easy to understand. The connection of each scientific topic to Divrei Chazal is impressive and really helps to drive home the beauty and complexity of each of Hashem's creations.

I am especially excited about the introduction of the latest textbook entitled *Anatomy and Physiology Madeb Family Edition.* This textbook goes into great depth, introducing students to the wondrous and miraculous workings of the human body in a very interesting, pleasant and relatable manner to inspire young minds. As a doctor, I am personally amazed at the quality of the content, illustrations and media that will keeps the students engaged and fascinated by the material.

May Hashem continue to give koach and bless Rabbi Lubin to continue to develop kosher teaching tools to enable students and teachers alike, to learn and teach science with a proper Jewish haskafah.

Sincerely,

Joseph Safdieh M.D.

Joseph J. Safdieh M.D.

Radiation Oncology Attending Physician
NY Health & Hospital Kings County Hospital
Clinical Assistant Professor
SUNY Downstate Medical Center

The Jewish Center for Science
3821 N. 51st Ave, Hollywood Fl 33021
Tel: 954.416.3305
Email: yaakovlubin@jewishcenterforscience.com

Dear Student,

Before we begin our journey exploring the wonders of Hashem found within our bodies—and discover millions of miracles constantly taking place—let's discuss how we should approach this endeavor from a Torah perspective.

"From my own flesh, I will see Hashem" (Iyov 19:26). We can explain this to mean: "Through my body—through the study of Niflaos HaBoreh—I can appreciate Hashem's wisdom." David HaMelech says, "The heavens declare the glory of Hashem and the sky speaks of His handiwork" (Tehillim 19:2). By studying the heavens, galaxies, planets, stars, atmosphere, and more, we can recognize Hashem's glory. The Rambam writes, "Rabbi Meir already wrote in a beraisa: Look at His deeds, because through this you will come to know [Hashem]" (Pe'er HaDor 57).

The Rambam (Hilchos Yesodei HaTorah 2:2) also explains that the way to achieve ahavas Hashem is by observing Hashem's wondrous creations—and recognizing that there's no comparison or end to His wisdom. Similarly, the sefer Chovos HaLevavos (Shaar HaBechinah and Shaar Avodas Hashem) teaches that through studying the wonders of the Creator and acknowledging that everything exists for our benefit, we'll want to become a true servant of Hashem, and want to do His will.

All aspects of science and nature reveal the wisdom of Hashem, Who created and continues to create everything. As David HaMelech states, "How great are Your deeds, Hashem. They were all made with wisdom" (Tehillim 104:24). Because of our limited vision and understanding, only a small fraction of this wisdom will ever be studied or known. Hashem's wisdom is far greater than we can ever fathom. This, too, is taught by David HaMelech: "Great is our Master and abundant in strength; there is no end to His understanding" (ibid. 147:5).

Unfortunately, those who study science often fail to recognize the incredible miracles right before their eyes. If we're going to study the brilliance of Hashem's beautiful, complex world, let's do it the way it's meant to be done—as a tool to reach ahavas Hashem and become a wholehearted eved Hashem, standing in awe of His ultimate wisdom and power.

It's my great honor and privilege to help guide you in exploring the wisdom of Hashem found within you—and my sincere hope that you enjoy the ride!

Rabbi Yaakov Lubin

Divrei Torah adapted from the writings of my rabbis and Chazal

TABLE OF CONTENTS

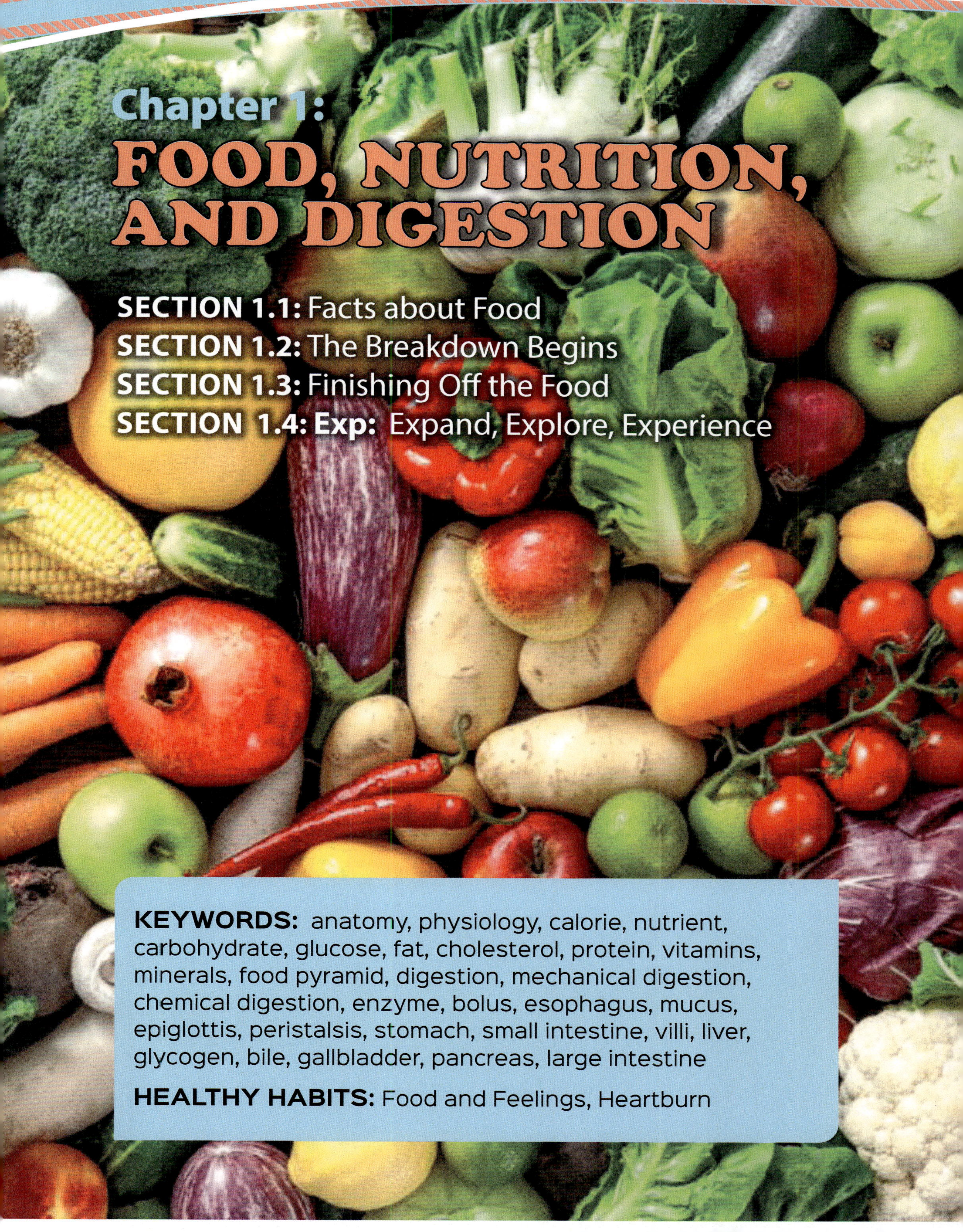

Chapter 1:

FOOD, NUTRITION, AND DIGESTION

SECTION 1.1: Facts about Food
SECTION 1.2: The Breakdown Begins
SECTION 1.3: Finishing Off the Food
SECTION 1.4: Exp: Expand, Explore, Experience

KEYWORDS: anatomy, physiology, calorie, nutrient, carbohydrate, glucose, fat, cholesterol, protein, vitamins, minerals, food pyramid, digestion, mechanical digestion, chemical digestion, enzyme, bolus, esophagus, mucus, epiglottis, peristalsis, stomach, small intestine, villi, liver, glycogen, bile, gallbladder, pancreas, large intestine

HEALTHY HABITS: Food and Feelings, Heartburn

SECTION 1.1: FACTS ABOUT FOOD

Just imagine yourself about to take a bite of chicken. That chicken is going to fuel your thoughts, movements, and the millions of internal activities going on each moment inside of you. If it didn't happen every day, it would be considered a fantasy land wonder that what started out as air, water, and light was transformed into a plant by way of photosynthesis. Then the plant was eaten by, absorbed into, and transformed into a chicken. And after that, the chicken was eaten by, absorbed into, and transformed into you!

This concept was told over to Rabbi Shlomo Diamond of Long Branch, New Jersey, by Rabbi Avigdor Miller, who emphasized how every time you eat meat, you should be thinking about how you are literally eating miracles. So, how does this incredible feat take place? How did Hashem design food so it could be used as fuel? And how can food be broken down, transformed, and reused by other organisms? These questions are going to be answered in this chapter, so get ready to be amazed—but before we begin, there are two important terms that we need know.

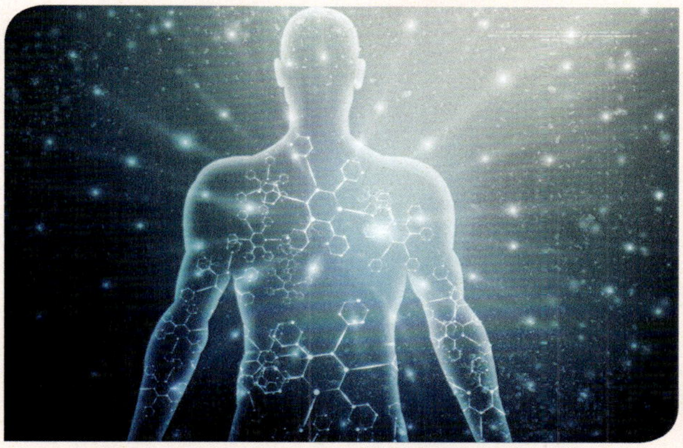

The study of the physical structures that make up your body is referred to as **anatomy** (uh-nat-uh-mee) and the study of what those structures are designed to do is called **physiology** (fiz-ee-ol-uh-jee). So, throughout this book, we will be exploring together the wonders of our Creator as seen in the anatomy and physiology of our own bodies (chapters 1–5), and the different structures found in the bodies of animals (chapters 6–8).

Why You Need Food

Why do you think you need to eat? One reason is that food provides the raw materials your body needs to grow and repair damaged cells and tissues. It is also the source of energy for everything you do, like walking, talking, and even thinking! This energy is stored in chemical bonds that hold food molecules together. Once the process of digestion has been completed (as described later in this chapter), those bonds break apart inside the cells of your body. As those bonds break, the energy that was holding the food molecules together is released and used to power all of the activities of your cells and body.

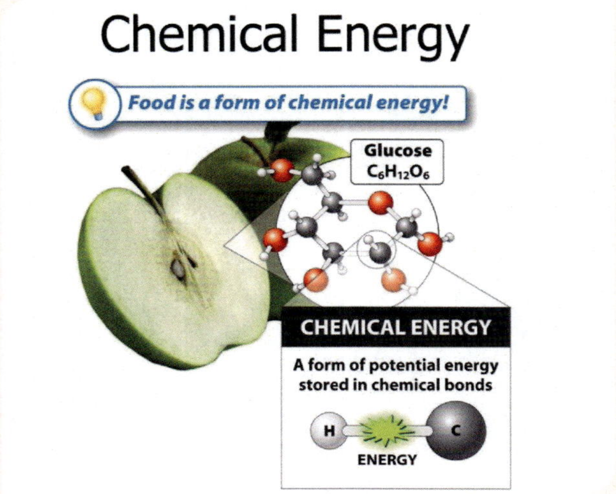

While the process of digestion takes a lot longer, the release of energy from the bonds of food molecules is very similar to what happens when you burn wood. The fire is coming from the release of energy that was holding the molecules of wood together. Hashem, in His infinite wisdom, designed the digestive process to take longer so the energy is released as we need it, little by little. That way, we don't have explosions taking place inside our cells, and the heat that is released is strong enough to provide us with power, yet a small enough quantity not to hurt us.

How much energy is inside of food? The amount of energy that can be released from a given portion of food is measured in calories. One **calorie** (kal-uh-ree) is defined as the amount of energy needed to raise the temperature of 1 gram of water by 1 degree Celsius. It is the unit of heat used to indicate the amount of energy that foods will produce in the human body.

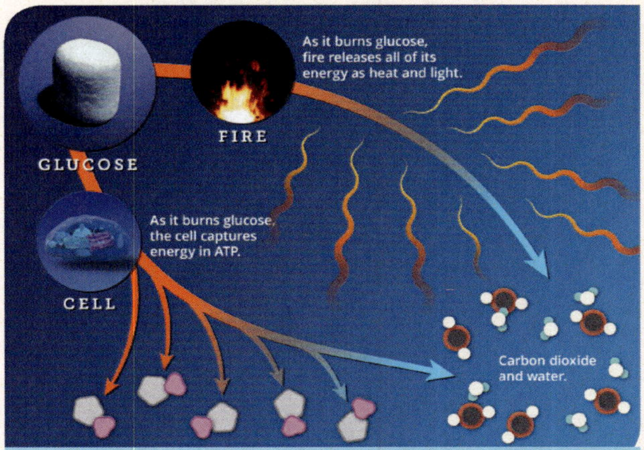

Both wood and marshmallows are made up of the same basic molecules and release heat when their bonds are broken.

Because most foods contain many thousands of calories, food labels use the term *Calorie* (with a capital *C*) to represent one kilocalorie (kcal), which is 1,000 calories. So a food label, such as one for a granola bar, that says there are 190 Calories is really saying that there are 190,000 calories—enough energy to raise 190 kilograms (419 pounds) of water by 1 degree Celsius.

Ingredients: Whole Grain Oats, Sugar, Canola Oil, Yellow Corn Flour, Honey, Soy Flour, Brown Sugar Syrup, Salt, Soy Lecithin, Baking Soda, Natural Flavor.
CONTAINS SOY; MAY CONTAIN PEANUT, ALMOND AND PECAN INGREDIENTS.

Nutrition Facts
Serving Size 2 bars (42g)
Servings Per Container 6

Amount Per Serving	2 bars		1 bar	
Calories	190		90	
Calories from Fat	60		30	
		%DV*		%DV*
Total Fat	6g	9%	3g	5%
Saturated Fat	0.5g	3%	0g	0%
Trans Fat	0g		0g	
Cholesterol	0mg	0%	0mg	0%
Sodium	160mg	7%	80mg	3%
Total Carbohydrate	29g	10%	15g	5%
Dietary Fiber	2g	8%	1g	4%
Sugars	12g		6g	
Protein	4g		2g	
Iron		4%		2%

Not a significant source of vitamin A, vitamin C and calcium.
*Percent Daily Values (DV) are based on a 2,000 calorie diet. Your daily values may be higher or lower depending on your calorie needs.

Now, if you read the label carefully, you will find a number of nutrients listed. A **nutrient** (noo-tree-uh-nt) is any substance in food that provides the raw materials and energy the body needs to carry out its processes. There are six basic types of nutrients that give you the energy you need and enable your body to function properly. They are: carbohydrates, fats, proteins, vitamins, minerals, and water.. Each one of these is crucial for your health and plays a different role in providing you with your needs.

Carbohydrates

The nutrients called **carbohydrates** (kahr-boh-hay-dreyts) are the main source of energy for the body and also provide the raw material for many of the structural components of cells. They are the most abundant biomolecule on earth. Biomolecules are those molecules produced by living organisms, as opposed to molecules coming from inanimate objects like rocks and metals. Carbohydrates are made up of the elements carbon, oxygen, and hydrogen.

Some foods that are rich in carbohydrates

Based on their chemical structures, carbohydrates are divided into two groups, called simple and complex carbohydrates. Simple carbohydrates are called "simple" because they contain only one or two units of carbohydrate molecules, as opposed to the many units that make up complex carbohydrates. Carbohydrates are actually made of sugar molecules. Simple carbohydrates have one molecule, and complex carbohydrates are made up of many sugar molecules in a chain. Simple sugars are commonly referred to as sugars, and include fructose, which makes honey, fruit, and fruit juices sweet; and lactose, which is the sugar found in dairy products such as milk, cheese, and yogurt.

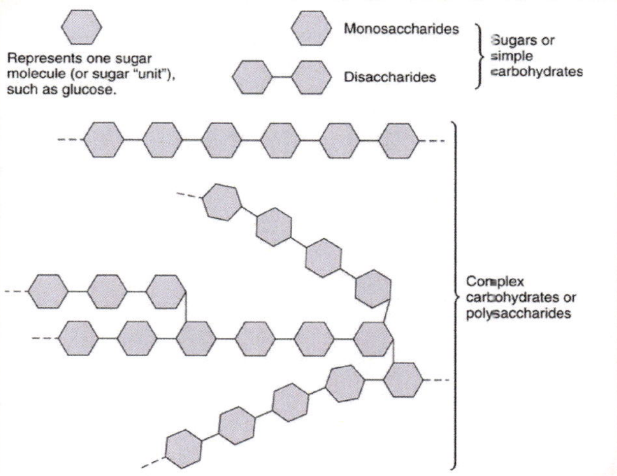

Sugars are also referred to as saccharides; those with one unit are called monosaccharides (mon-uh-sak-uh-rahyds), those with two are called disaccharides (dahy-sak-uh-rahyds), and those with more than two are called polysaccharides (poly-sak-uh-rahyds) or complex carbohydrates.

Of the various types of sugars out there, the one called **glucose** (gloo-kohs) is the main source of energy for your body's cells. Most foods don't actually contain much glucose, but the other types of sugars are broken down into glucose, which is smaller and easier for the body to turn into energy you can use.

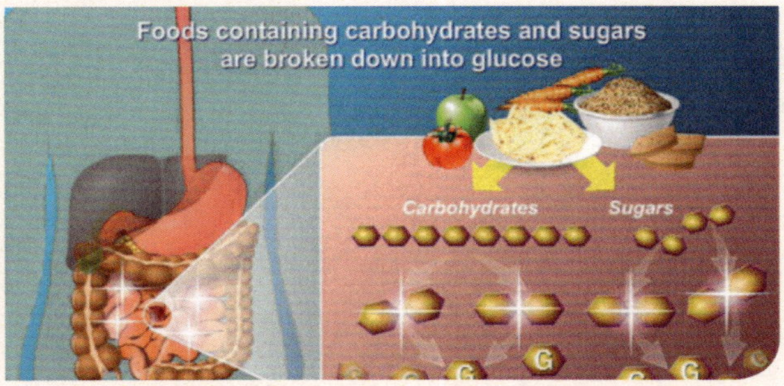

As we mentioned above, complex carbohydrates are made up of many sugar molecules linked together in a chain. Plants are made up of an abundance of them, referred to as starches. Starches are found in foods such as potatoes, rice, and wheat. For the body to release and utilize the energy stored in starches, the starches must first be broken down into their individual sugar molecule components through digestion.

Another type of complex carbohydrate is called fiber. Fiber is different than the other types of carbohydrates because it cannot be broken down into individual sugar molecules that your body can use. Despite the fact that it cannot be used for energy, fiber is very important to the maintenance of your health. It passes through your digestive system whole and helps move all the other stuff along smoothly, greatly assisting the digestive process. Fiber provides other health benefits as well, such as lowering the risk of diabetes, heart disease, and some types of cancer.

Fats

Fats are perhaps the most underappreciated nutrient out there. While it is true that one should avoid eating too much fatty food since it can cause people to become overweight and lead to other health issues, nevertheless, fats are still extremely important to your well-being.

Some foods that are rich in fat

One of the roles that fat plays is to help form the fatty tissue that serves as a sort of cushion around your internal organs, supporting and protecting them. Fatty tissue also helps insulate your body from the cold, and fat molecules make up the majority of cell membranes. This is all in addition to the fact that fat is a major source of fuel for your body, as it contains more than twice the amount of energy per gram as carbohydrates or proteins.

Types of Fats

The key to knowing which fats are healthy and which ones should not be eaten often is knowing the difference between saturated and unsaturated fats. All fats have the same basic elements of carbon, hydrogen, and oxygen in long tail-like structures.

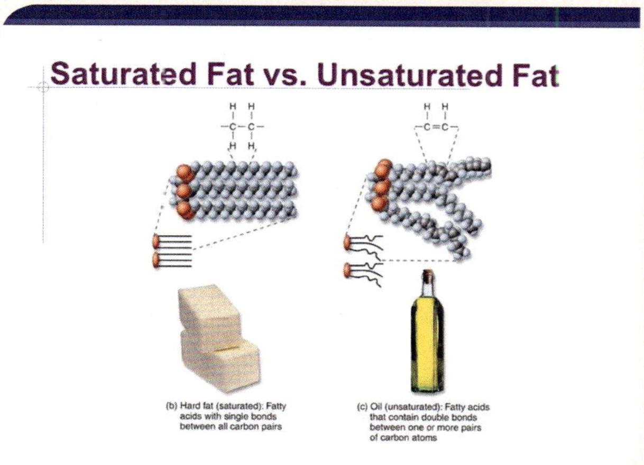

The reason why some are called saturated and others unsaturated has to do with the number of hydrogen atoms they have in their molecular structure. All living things contain carbon atoms that make up the "backbone" of the molecule. Carbon atoms can attach to four other atoms, which is more than most. Because it has so many points of attachment, there are several different shapes and types of molecules that are built using carbon atoms. Many children's toys, such as Clics, use the same principle by putting four points of attachment at the ends of the plastic pieces, enabling a great variety of structures to be built. The versatility of carbon gives rise to the abundant variety of biomolecules, the molecules of life.

Saturated fats have the maximum amount of hydrogen atoms possible attached to their carbon backbone, which generally results in their being solid at room temperature. Unsaturated fats are missing one or two hydrogens in their chains. The "missing link" puts a kink in the "tail." This results in fats that are liquid at room temperature, like olive and vegetable oils.

Unsaturated fats are generally much healthier than saturated ones. They are found in foods like vegetable oil, nuts, and olives, and provide an important source of nutrition. Saturated fats are found in meat, dairy products, and packaged foods, like prepackaged cookies and potato chips. Eating excessive amounts of saturated fats can lead to serious health issues.

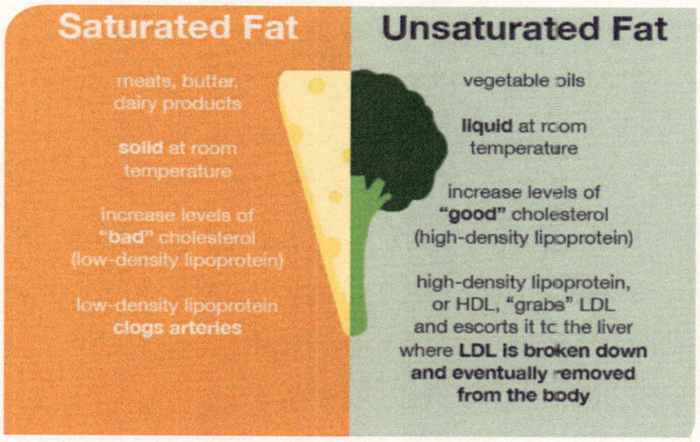

One of the most common is heart disease that is caused by a buildup of **cholesterol** (kuh-les-tuh-rohl), also known as a LDL (low-density lipoprotein). Cholesterol is a waxy substance found in animal products. Like other fats, cholesterol is an important part of your body's cells. However, unlike other fats, your liver produces sufficient amounts of it, so you don't need more from food.

Because of its waxy consistency, an excess of cholesterol can cause a buildup of plaque

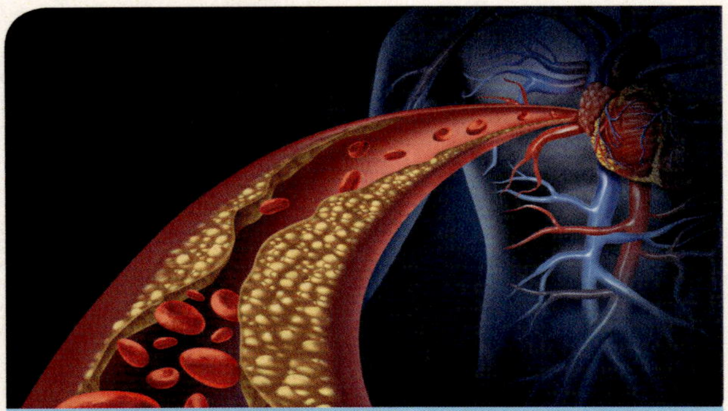

A buildup of cholesterol can clog up arteries, which bring oxygen-carrying blood to the cells of the body.

in the arteries, which are blood vessels that bring oxygen and nutrients to the cells throughout your body. Over time, the plaque can get so thick that it can clog the arteries, making it difficult or impossible to get the necessary oxygen and nutrients to the cells in your body. When this happens in vital organs like the heart, it can cause a heart attack, which is what happens when the oxygen supply is cut off to the heart. In the event that arteries leading to the brain become clogged, it can result in a stroke, which happens when brain cells do not receive the oxygen they need.

Proteins

Some foods that are rich in protein

Proteins are actually much different than carbohydrates and fats in that they are a much less important source of energy. While they can be used to fuel your body's activities if necessary, they are most often used for tissue growth and repair, and play an important part in chemical reactions within cells.

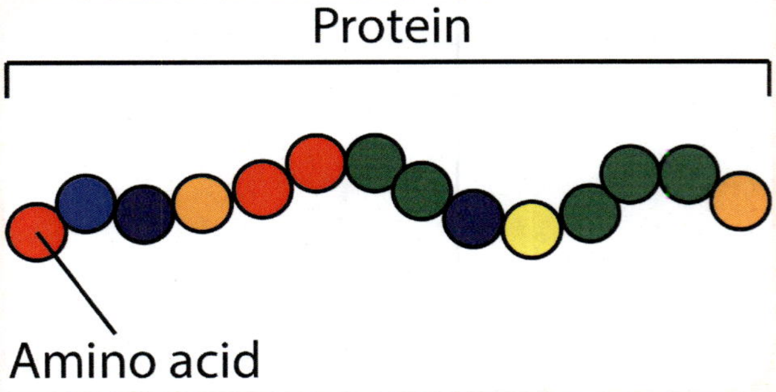

Proteins are made up of different combinations of small units called amino acids. There are 20 different amino acids, and your body can produce about half of them. The other half, called essential amino acids, you must get from the foods you eat. After you eat proteins, they are broken down into their amino acid components, which are then used to build the various types of proteins used by your body to do many of the necessary functions of life, such as growing muscles and making the microscopic machinery used by cells.

There are two different types of proteins: complete and incomplete. Foods from animals, for example meat and eggs, are sources of complete proteins. That means they contain all the essential amino acids. Proteins that come from plants, such as beans, grains, and nuts, are each missing one or more of the essential amino acids, although not necessarily the same ones.

These foods are therefore called incomplete proteins. The incomplete nutrition provided by plant-based food often poses a challenge to vegetarians who, by definition, choose not to eat any meat products. In order for vegetarians to get all the amino acids they need, they must be careful to eat a wide enough range of plant products so that combined they will consume all the amino acids they need.

If you take a minute to think about these three types of fuel we use to fill up our body's gas tank every day, it is pretty amazing. There are little packets of energy locked up in the molecular bonds that hold together the food we eat. After being broken down into microscopic parts, the fuel is burned in the numerous "engines" found in every one of the 100 billion cells of your body. And what's the most impressive part? When the tank is running low, there is an automatic warning system that tells you it's time to refill the tank—you get hungry!

Vitamins

While food definitely provides the fuel that makes you go, and the raw materials needed to replace old or damaged cells, it contains much more than that. Each day, there are literally trillions of chemical reactions and complicated processes going on inside of you. To facilitate these processes, Hashem made vitamins and minerals. Although they are only found in small amounts and do not provide the body with energy or raw materials, they are essential to keeping the factories of life within you running smoothly.

You have probably taken vitamins before and know they are important for your health, but what exactly are they? **Vitamins** (vahy-tuh-mins) are sort of like helper molecules for a variety of chemical processes that take place inside of your body. For example, vitamin K helps with blood clotting when you get a cut, several types of B vitamins help release energy from food, and vitamin A helps keep cells healthy as well as protects your vision. There are 13 known vitamins and most of them cannot be produced by the body. Instead, they are ideally absorbed when we eat healthy foods. Nevertheless, many people choose to take pill or capsule supplements to ensure that they are getting all the vitamins they need.

Vitamins can be divided into two groups: water-soluble and fat-soluble. Water-soluble vitamins dissolve in water and your body cannot store them for long periods of time. That means that you can't really take too many of them, but it also means that you need to make sure to get enough of them on a regular basis. Vitamins C and B are in this category.

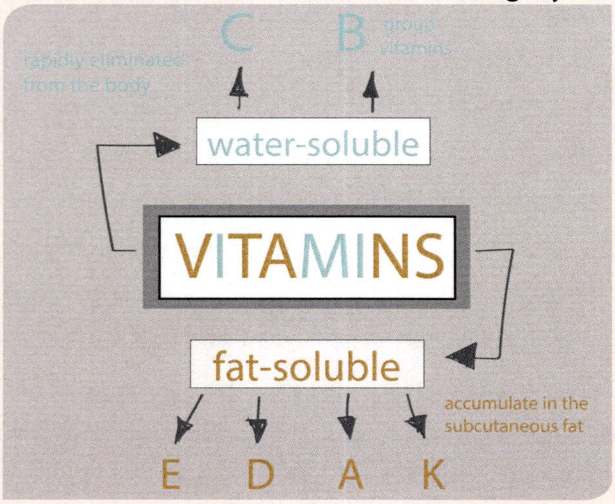

Fat-soluble vitamins dissolve in fat, not water. They are stored in the fatty tissues in the body and include vitamins A, D, E, and K. The fact that they can be stored means you could technically take too many of them, which could, in turn, cause health problems, but most people nowadays need to focus on getting more of them to stay healthy and feel good.

Minerals

Nutrients that are not made from living things are called **minerals** and include iron, calcium, and magnesium. Like most vitamins, your body does not produce minerals and they must be obtained from the food you eat. Minerals are needed to carry out vital chemical processes taking place within your body, for example when calcium helps build strong bones and magnesium aids in nerve and muscle function.

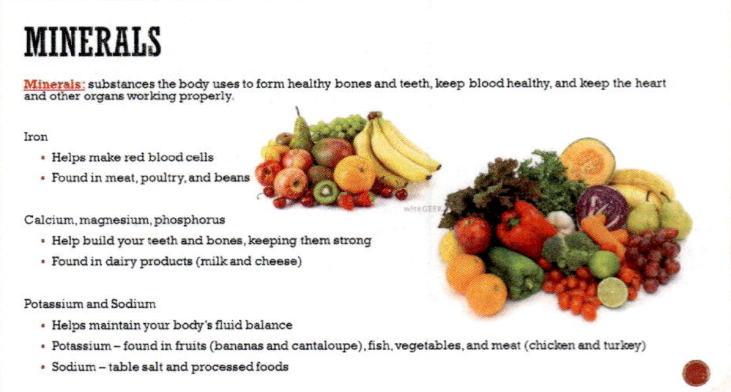

Water

Last but certainly not least is perhaps the most essential nutrient of all: water. While someone could probably survive for a week or two without food, people cannot go more than a few days without water. The majority of your weight is actually made up of water (about 65%). Water is so important because all of the body's processes, such as the breaking down of food, can only take place in the presence of water.

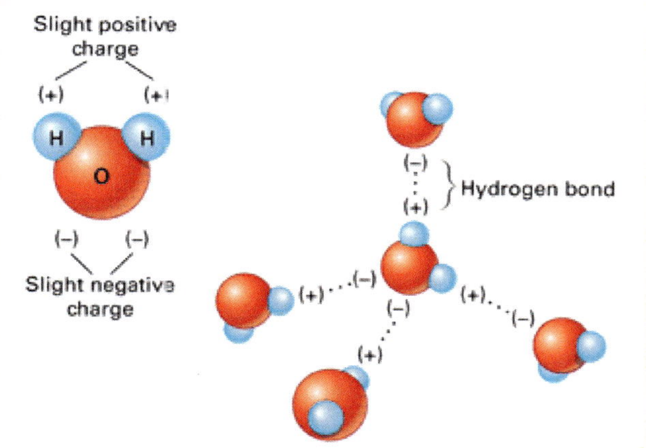

The secret to water's success lies in its exquisite design. Water molecules are made in such a way that they stick to one another. If you don't believe me, try filling a cup of water a little bit above the rim. The reason it doesn't spill over is because the individual water molecules pull on each other in very weak and short-lived bonds that scientists call hydrogen bonds. Although each bond is frail and doesn't last long, at any given time most of the water molecules are holding on to each other, thus preventing the water from spilling over.

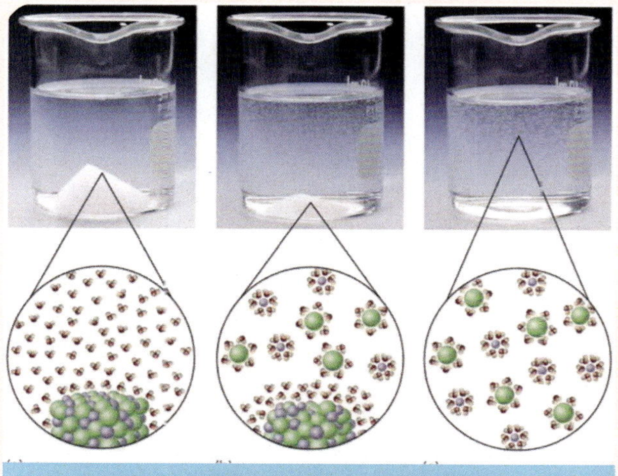

How water works its wonders!

Water molecules don't just form bonds with each other; they actually bind to just about everything they come in contact with. When something such as salt falls into water, the water molecules grab onto and pull apart the individual atoms that it is made of. In a short time, water can dissolve many of the substances needed for life. If not for water, the nutrients you need to stay healthy would be too big for the microscopic cells of your body to use. Water helps break the nutrients down, making them useful and easily transportable to all parts of your body. In fact, water's ability to break other substances down is so remarkable that scientists refer to it as the "universal solvent." A solvent is a substance that can break down other things, and water is better at doing this than anything else on earth!

Staying Healthy

There are different opinions out there as to the best foods to eat and how to stay fit, but most doctors agree on some general guidelines. Healthy eating is often taught using a **food pyramid**, which is a pyramid-shaped diagram showing the recommended daily servings from different food groups.

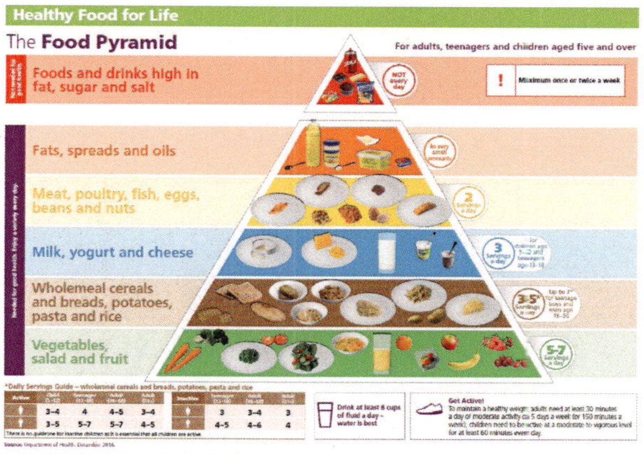

In addition to eating healthy, it is very important to get enough water and exercise regularly. Drinking 8 cups (about 2 liters) of water should keep you hydrated (supplied with water), and adults should exercise at least 30 minutes a day, while children need at least an hour.

Seeing that the Torah teaches us, "*Venishmartem meod lenafshoseichem*—Be very careful about your lives" (*Devarim* 4:15), it is very important that we do our best to take care of our health. Author Jack Botwinik, in the Orthodox Union's magazine *Jewish Action* (Spring 2008) quoted the Rambam (Rabbi Moses Maimonides), from his work entitled *Hanhagos Habriyos* (The Regimen of Healthcare): "If a person cared for himself the way he cares for his horse, he would avoid many serious illnesses. You won't find a person who gives his horse too much fodder. But he himself eats to excess. He makes sure his animal gets proper exercise to keep it healthy. But when it comes to himself, he neglects exercise even though this is a fundamental principle in health maintenance and in the prevention of most illnesses."

So, the next time you have a break and choose to play a game of soccer, basketball, or ride a bike, in addition to having fun, make sure you have in mind to fulfill a mitzvah—taking proper care of your body!

HEALTHY HABITS:

Food and Feelings

While all doctors and nutritionists agree that eating a diet that consists of fruits and vegetables is fundamental to one's health and well-being, not many discuss the affects they can have on one's emotional health. More and more studies have been coming out from psychologists and therapists that show a direct connection between eating a healthy diet and feeling good emotionally. In one case, Rabbi Yosef Nemiroff, a psychotherapist and nutritionist, reports how he used food to successfully treat a client with OCD (obsessive compulsive disorder), a condition where someone has constant negative thoughts and behaviors that they can't stop. After much work and little progress using typical methods of treatment, Rabbi Nemiroff suggested that the patient look into his diet. After expressing doubts that diet was even connected to his mental health condition, the patient reluctantly agreed to add more fruits and vegetables to his diet. Two weeks later, when he was still eating more vegetables, he had an amazing breakthrough. He was much more relaxed and for the first time in his life, his thoughts and feelings were more positive and flowed more normally. The truth is that according to studies on the relationship between food and feelings, most people live happier and more emotionally stable lives when eating a healthy diet. So...try eating more fruits and vegetables and let the good feelings flow!

Foods for Every Mood

Distracted?

Whole grains, bananas and milk contain tryptophan, a **natural sedative**.

Pulling an all nighter?

Fish and eggs are known to **enhance memory**.

Anxious?

Chia seeds, fish oil, walnuts and salmon have omega-3, which promotes **calm feelings**.

Tired?

Dark chocolate, green tea and avocados all contain **natural mood balancers** such as serotonin, tryptophan and theanine.

SECTION 1.1 REVIEW

1. What are the two types of carbohydrates? Give an example of each one.

____

2. What are two types of fats? Give an example of a food each one is found in.

____

3. Which of the two fats is less healthy? What type of health issues can arise from eating too much of them?

____

4. What types of proteins do you need to eat to ensure you are getting all of the essential amino acids?

____

5. What do vitamins and minerals have in common, and how are they different?

____

USE YOUR MIND

If you were stranded on a deserted island and could have only one source of energy for a week—carbohydrates, fats, or proteins—which one would you choose and why?

____

____

SECTION 1.2:
THE BREAKDOWN BEGINS

This section explains the mystery of how food is broken down into molecules that are used to power your body and obtain the nutrients you need.

Now that we know what types of foods you need to eat, the next adventure is to discover how food is broken down. Follow its journey throughout your body, and find out why you really are what you eat!

The process of breaking foods down into small nutrient molecules the body can use is called **digestion**. The series of organs that food passes through in this process is called the digestive tract or GI (gastrointestinal) tract. It may surprise you to hear that while most people think digestion starts when you put food into your mouth, it actually starts with your eyes and nose.

Sometimes just thinking about your favorite food can make your mouth water.

If you have ever had some warm, delicious-smelling food put onto the table, you may have noticed your mouth starting to water. The sight and smell of foods as well as chewing actually send triggers the brain to start the digestive process. Your body produces more saliva in the mouth and digestive juices in the stomach that are necessary for breaking down food.

The nose knows... what you should and should not eat!

In addition to its role in digestion, the nose acts like a security guard to ensure spoiled foods don't enter our bodies. Just like security guards prevent dangerous people from entering buildings, by inspecting their credentials before allowing them in, so too Hashem gave the nose the ability to identify unpleasant odors that come from foods that are spoiled and unfit for human consumption.

The eyes are often deceived and can't tell the dangers lurking inside spoiled foods, but the nose knows when something smells off and is revolted by the foul odor. The fact that the nose extends out over the mouth ensures that it will get a whiff of what is about to enter and can advise the would-be eater not to allow that food entrance. This brings us tremendous benefit, and is a very important function for which we should thank Hashem.[1]

Mouth

Once food is inside the mouth, the two types of digestion begin. The first kind is called mechanical digestion and begins with your teeth. **Mechanical digestion** refers to the physical breakdown of large pieces of food into smaller pieces through the chewing of food in your mouth and then the churning of the food in your stomach. The second type of digestion is called **chemical digestion** and involves the breakdown of foods into small molecules the body can use by way of chemicals called **enzymes** (en-zahyms).

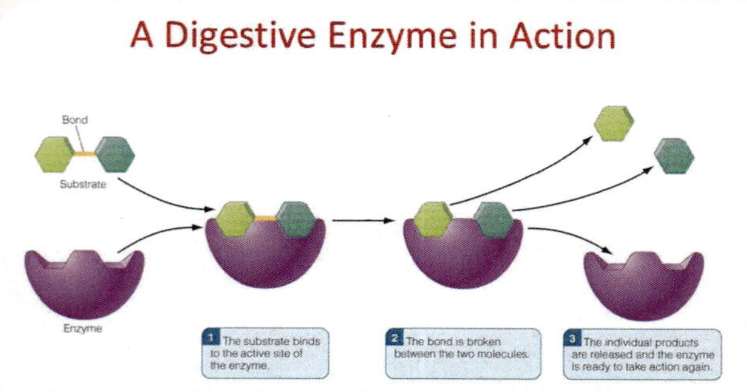

The substances enzymes work with are called substrates. The unique shapes the substrates fit into on the enzymes are called active sites.

1 Adapted from Ohr Avigdor by Rabbi Avigdor Miller.

Enzymes are a type of protein designed to speed up chemical reactions in living things. Chemical digestion begins in the mouth when an enzyme in your saliva called amylase (am-uh-leys) begins the chemical breakdown of foods containing complex carbohydrates (starches) into individual sugar molecules.

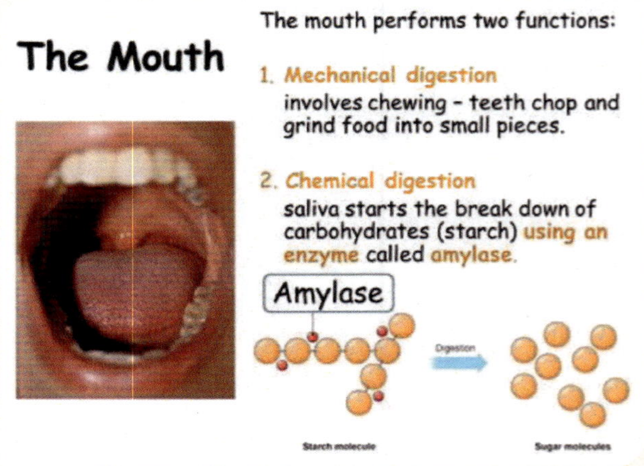

In addition to providing digestive enzymes, saliva aids the digestive process by moistening your food so that it is easier to chew and can go down into your stomach smoothly when swallowed. This ensures that the sharp edges of things like potato chips that weren't chewed properly or hard foods like nuts don't cause damage on their way down. The lubricated food that gets mushed into a ball in the back of your mouth before it is swallowed is called a **bolus** (boh-luhs).

It is obvious how your teeth grind up and break down food, but how does an enzyme like amylase do it? Inside your body, there are many different types of enzymes that aid in virtually every chemical process that takes place. Enzymes work their wonders by way of specially shaped "pockets" on their surface, and the one specific type of molecule that they were made to react with fits into it perfectly. This ensures that enzymes don't go around doing the wrong thing by causing chemical reactions when they aren't supposed to. They only react with the substance that fits in their pocket—the substance they were designed to partner with, and in this way carry out the millions of reactions taking place within you every moment.

Esophagus

Once food has been mashed up by your teeth, lubricated by saliva, and squished into a ball, it is ready to begin its descent into your stomach. There is a stretchy pipe made out of muscles about 10 inches (25 centimeters) long that connects your mouth to your stomach, and it's called the **esophagus** (ih-sof-uh-guhs). The whole esophagus is lined with **mucus** (myoo-kuhs), which is a thick, slippery substance that helps food move down the esophagus and into the stomach smoothly.

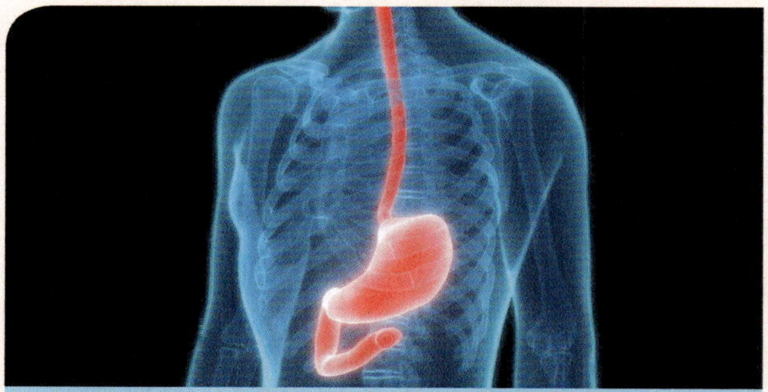

Food traveling down the esophagus on its way to the stomach

In order for food to make it safely to the esophagus, there is a potentially dangerous problem that must be overcome. There are actually two separate openings at the back of your mouth. One leads to your windpipe, which carries air into your lungs (see chapter 2 for the details), and the other is the esophagus pipe that leads to your stomach. If food goes down the wrong pipe, it can prevent the passage of air and cause a person to choke. To bypass this problem, Hashem made a flap of skin called the **epiglottis** (ep-i-glot-is) that automatically seals off your windpipe every time you swallow, thereby preventing food from going the wrong way.

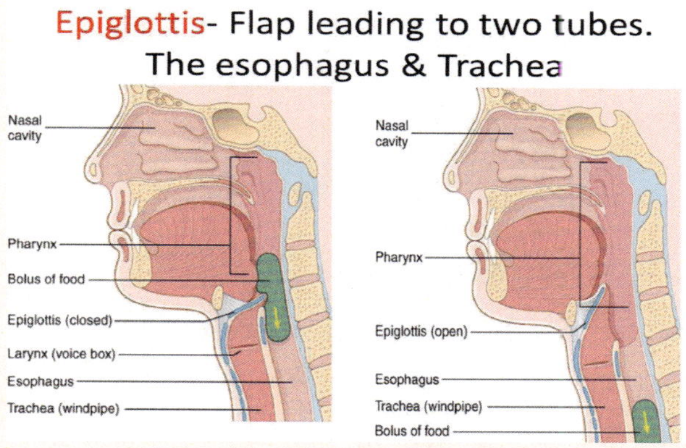

It is for this reason the *Shulchan Aruch* (*Orach Chaim* 170:1) states that one should not talk while eating, as it can open up the epiglottis and result in food going down the wrong pipe, causing a person to choke.

Now, the food doesn't just fall down into your stomach. Rather, it moves by way of a series of muscle contractions that push the food toward the stomach. While you can control whether you swallow or not, once your body starts this process, the wave of muscle contractions continues on its own until the food has moved all the way into the stomach. The food moves not just through the esophagus, but all along the digestive tract in this manner. These muscle contractions are referred to as **peristalsis** (per-uh-stal-sis) and ensure that food stays moving in one direction—the right one!

PERISTALSIS

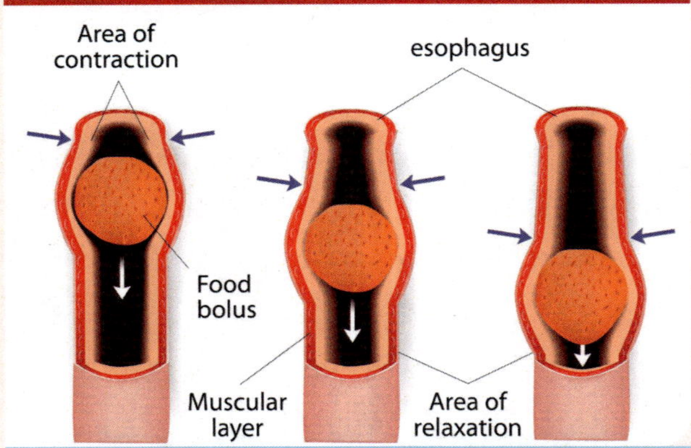

It only takes about 10 seconds for food to travel from the esophagus to the stomach by way of peristalsis.

Stomach

I am sure you are well aware that you have a stomach, but do you really know what your stomach looks like and what it does? The **stomach** is a J-shaped muscular pouch that receives the food coming from the esophagus and performs most of the mechanical digestion of your food. Food is kneaded together over and over in a churning motion made by three strong layers of muscle to ensure it turns into a thick liquid. Depending on the type and size of the meal, the process of pulverizing the food in the stomach takes between 2 and 6 hours before it is ready to move on to its next destination—the small intestine.

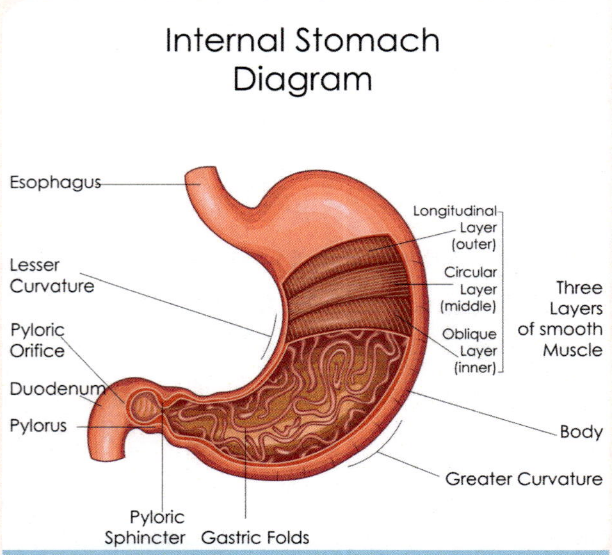

There are three layers of muscle in the stomach, turning in different directions to crunch foods from all angles.

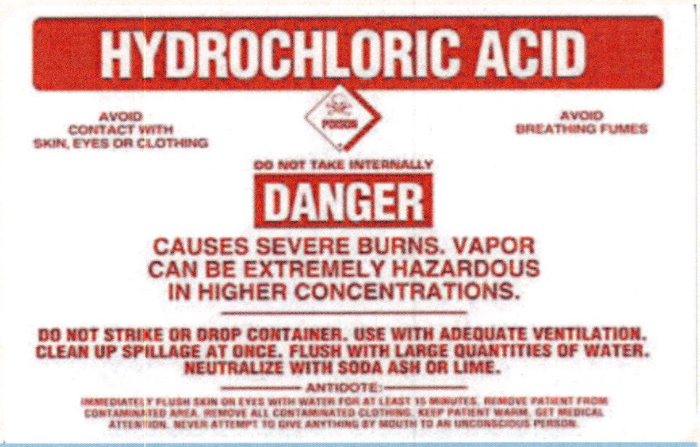

Only Hashem knows how to put a substance that is lethal to drink right inside your stomach!

In addition to the mechanical digestion that takes place in the stomach, an enzyme called pepsin begins the chemical digestion of proteins, breaking them down into short chains of amino acids. In addition to pepsin, the digestive juices of the stomach contain hydrochloric acid, a very powerful acid that eats away at proteins, thereby helping to break them down. This acid is so strong that if you put your finger in it, your skin would get burned.

So you are probably wondering, if this acid is so strong that it can digest the proteins in the tough meat that you eat, why doesn't it digest the inside of your stomach—*your* meat? The answer involves another one of the miracles taking place inside of you every day. Hashem designed the lining of your stomach with special cells that secrete a thick layer of mucus to protect you from this powerful acid. In addition, the mucus cells are rapidly reproduced since the harsh acidic environment wears the cells out quickly.

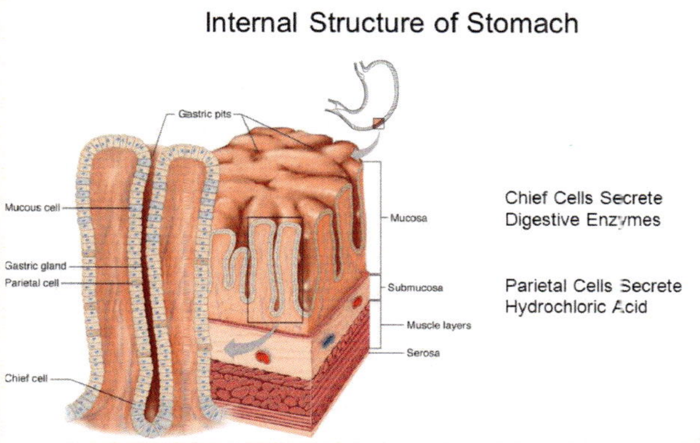

The hydrochloric acid in your stomach also helps in another important way. Despite mankind's best efforts to keep foods clean and fresh, foods often contain bacteria that have the potential to make you ill. The acid is so strong that it kills many of these bacteria, so when the food eventually goes into your blood, it is free of harmful bacteria.

A Hole in History

The role that hydrochloric acid and digestive enzymes have in digesting food was not known to the scientific community until the early 1800s. The great discovery came through a very unexpected source— a gunshot wound in the stomach. In 1822, a man named Alexis St. Martin was shot in the stomach. His doctor, Dr. William Beaumont, actually expected he would die. Despite his dire prediction, however, Dr. Beaumont did his best to treat the wound. St. Martin lived, but the gunshot hole in his stomach never fully closed.

Dr. Beaumont considered the hole in St. Martin's stomach a unique opportunity to study digestion. He conducted many experiments by tying a piece of food to a string and inserting it into St. Martin's stomach. After various amounts of time, he would record how well the food was digested as well as analyze the gastric juices (juices secreted by the stomach that were covering the food). This was the first time scientists realized that digestion was primarily a chemical process and not a mechanical one.

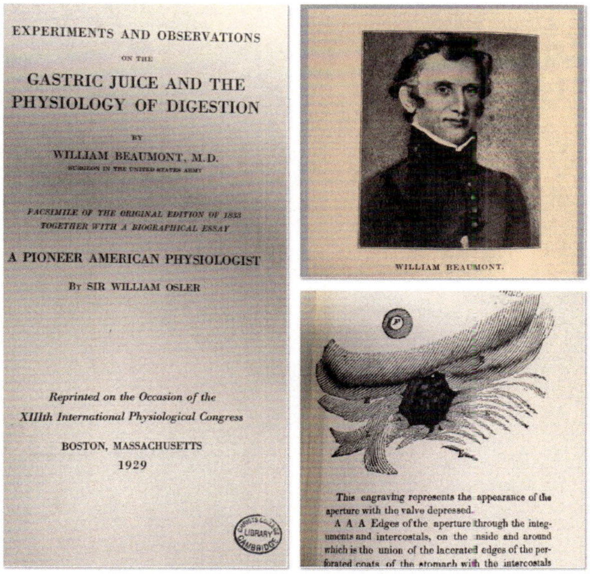

Dr. William Beaumont's work explaining digestion, and the picture he drew of the hole in Alex St. Martin's stomach

HEALTHY HABITS:
Heartburn

Heartburn is a common painful sensation that occurs in the chest, near the heart, but most people don't really understand why it happens or its connection to the esophagus. The esophagus has a ring of muscles at the top and bottom. If the bottom ring of muscles doesn't keep the lower part of the esophagus tightly closed between swallows, acid from the stomach can come back up into the esophagus. When the bottom ring of muscles does not close properly, or when it opens too often, acid from the stomach comes up and causes a burning feeling that can last anywhere from a few minutes to a few hours. Heartburn can be avoided with the practice of a few healthy habits that include: not overeating; avoiding fatty, spicy, or greasy foods; and not lying down immediately after eating. In addition, for those suffering from chronic heartburn, doctors suggest avoiding certain foods such as: fatty or fried foods, coffee, onions, tomatoes, soda, chocolate, peppermint, and citrus foods (oranges, lemons, etc.).

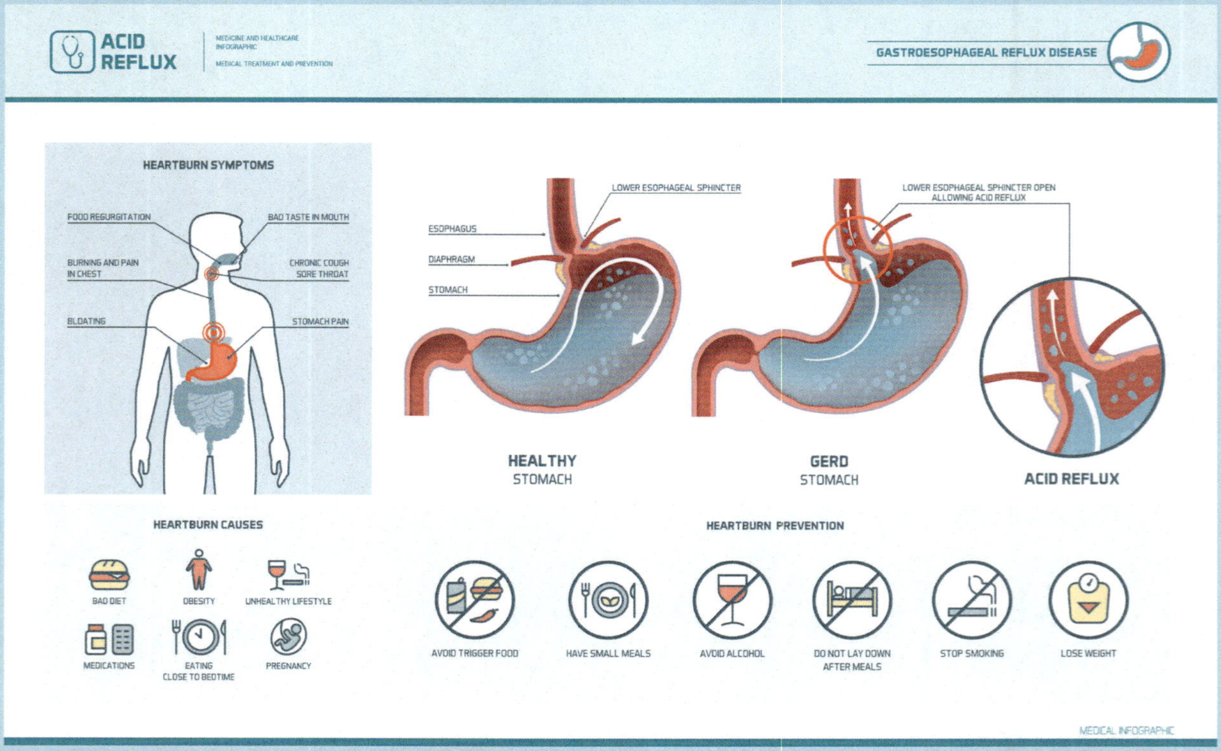

SECTION 1.2 REVIEW

1. What is digestion and where does it begin?

____

2. Explain the two different types of digestion that take place in the body.

____

3. After swallowing, how does food move down the esophagus and digestive tract?

____

4. What prevents food from going down the wrong pipe—meaning, down the windpipe instead of the esophagus?

____

5. In what ways does food get digested in the stomach?

____

USE YOUR MIND

During Dr. Beaumont's experiments, which types of food were more digested when he pulled them out of Alexis St. Martin's stomach and which ones were less digested?

____

____

SECTION 1.3
FINISHING OFF THE FOOD

Before nutrients actually go into the bloodstream to be distributed throughout the body, they must pass through the stomach and move on to the small intestine. There, the remainder of the food gets digested and nutrients are passed into the bloodstream.

Small Intestine

Once the stomach completes its mechanical digestion stage, the resulting thick nutrient-rich soup referred to as chyme (kahym) is passed on to the next stage of the digestive process in the small intestine. However, not all the food passes at once. Small amounts at a time are let in by way of a valve (a door-like structure) called the pyloric sphincter (pahy-lawr-ic sfingk-ter), which opens and closes to ensure that the food passes slowly enough to enable maximum digestion and absorption.

PYLORIC SPHINCTER OF THE STOMACH

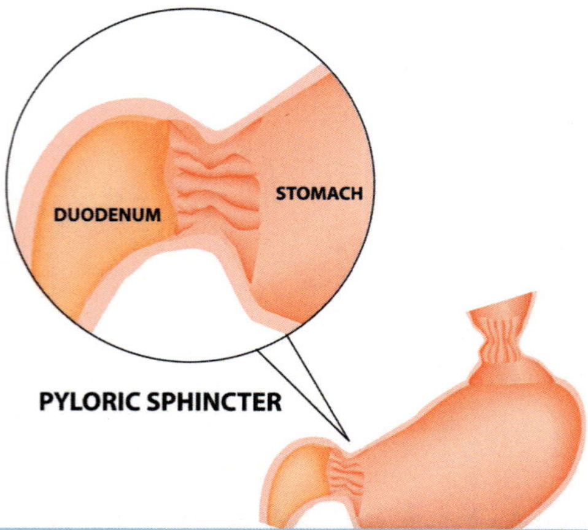

The duodenum is the first segment of the small intestine. The pyloric sphincter regulates how much food enters it at a time.

While the name may make you think we are talking about a small organ, the **small intestine** (in-tes-tin) is actually about 2 inches (5 centimeters) round and 22 feet (6.7 meters) long—that makes it longer than a pickup truck! Although the mouth and stomach do a good job of mechanically digesting food, it is in the small intestine that most of the chemical digestion and all absorption of nutrients take place.

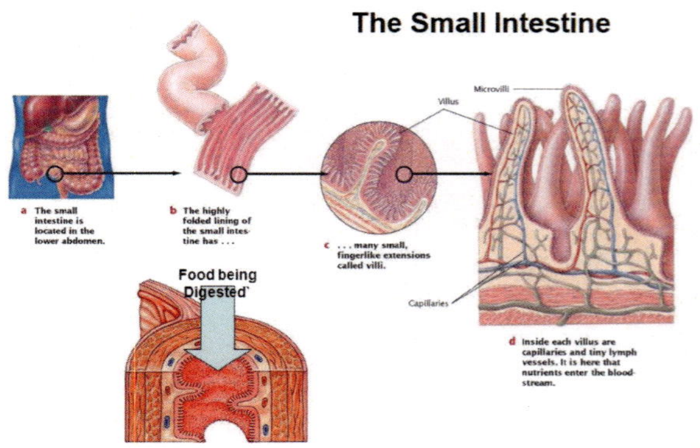

Since none of the fats and only a portion of the carbohydrates and proteins you ate were digested in your mouth and stomach, enzymes capable of digesting fats, carbohydrates, and proteins are found in the small intestine. This is important since it is here that the nutrients from your food are absorbed into the bloodstream and then transported to cells throughout your body. To facilitate the absorption of these small nutrient molecules,

Hashem designed the inner surface, or lining, of the small intestine with millions of folds or bumps called **villi** (vil-ahy) (singular: villus [vil-uhs]). In fact, there are so many villi that if stretched out, they would be around the size of a tennis court! The villi greatly increase the surface area and enable many more nutrients to be absorbed than would be possible if it were flat.

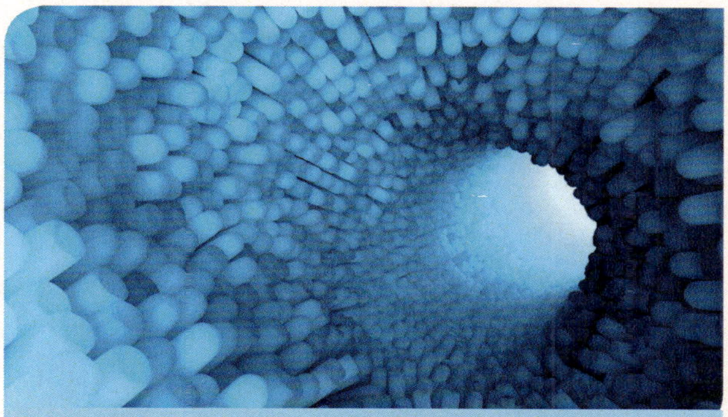

3D illustration of the villi located inside the small intestine.

Villi absorb nutrient molecules by way of tiny blood vessels that pass through the center of each villus. After chemical digestion has been completed, the nutrient molecules are small enough to pass from cells on the surface of the villi into blood vessels. Once absorbed into the blood, the nutrients are taken throughout the body for the cells to use.

Liver and Pancreas

Two other organs, the liver and pancreas, help the small intestine accomplish its task of completing the full breakdown of food. They both deliver substances that help with the chemical breakdown by way of small tubes that lead directly into the small intestine. The first is the **liver**, which is the largest organ in the body and plays an important part in many body processes in addition to digestion.

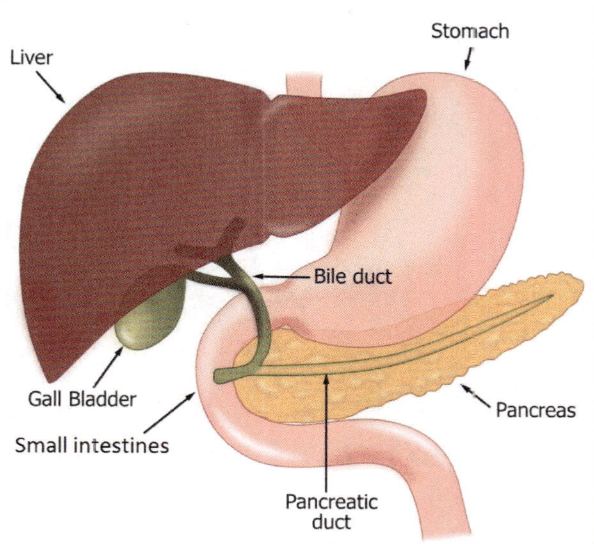

Over 500 vital functions have been identified with the liver. It is so remarkable that it has been called the body's "miracle laboratory." One of the helpful things it does is remove toxins (substances in the body that are actually like poisons) from your blood. The liver also receives and cleans blood of foreign particles directly following the digestive process that takes place in the small intestine. Before going out to the rest of the body, the nutrient-rich blood still needs some processing before going off to supply the cells of your body with the materials they need. The first stop of blood following digestion is the liver, which modifies many of the substance into forms that the rest of the body can use. The waste, or stuff your body doesn't need, is then carried back to the large intestine, where it is prepared for elimination from the body.

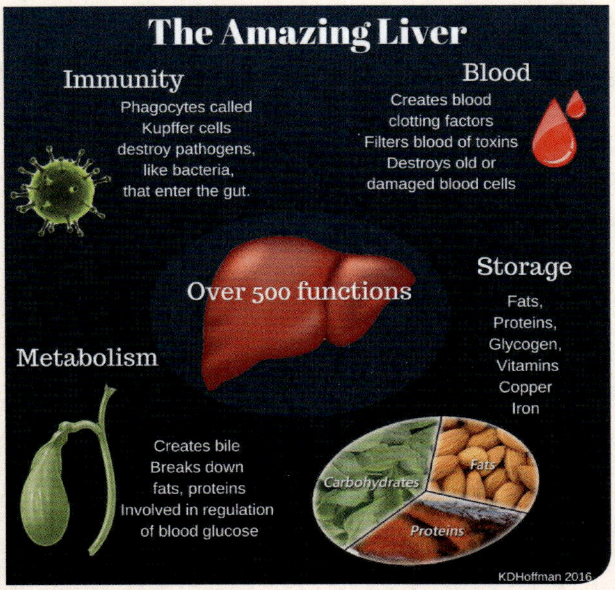

The liver also builds proteins and makes a complex carbohydrate called **glycogen** (glay-kuh-jen), which stores sugar molecules in long and complex chains. As we learned earlier, sugars can only be used by cells when they are single molecules, so glycogen stores them for when food is scarce. When needed, the liver "taps" the reserves by breaking down glycogen into sugars that can be used.

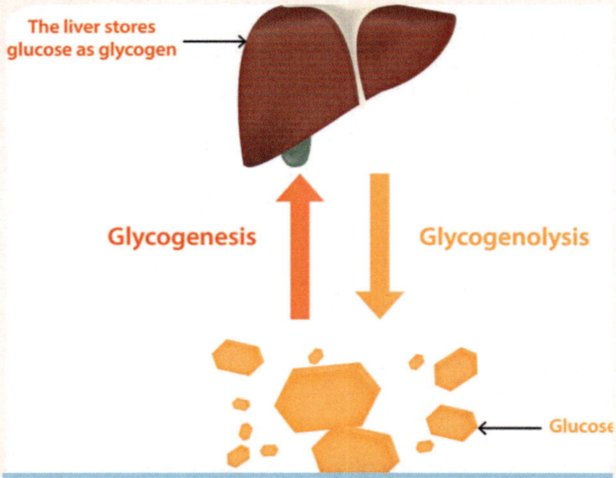

The liver converts glucose to glycogen, and then the reverse, according to the needs of the body.

So the liver continually converts glucose to glycogen and then glycogen to glucose according to the needs of the body. This is a great *chesed* in Hashem's design of our bodies. The presence of glycogen helps us store food in order to survive fast days, or even a regular school day when we forgot our lunch on the kitchen counter!

To facilitate digestion, the liver also produces a substance called **bile** (bahyl), a yellowish-green fluid that breaks down fat particles. Bile is made in the liver, but flows into a small organ called the **gallbladder** (gawl-blad-er), where it is stored. After you eat, bile passes through a small tube into the small intestine to help digest your meal. It is not an enzyme, so it doesn't break down food using chemical reactions. But it does aid in digestion by physically breaking up large fats into smaller particles that are easier for digestive enzymes to pick apart. Ever try to wash chicken grease off your dinner plate? Water alone literally doesn't cut it, but soaking the plate or scrubbing it with soapy water breaks the fats apart and does the trick. Bile acts a lot like dishwashing liquid or laundry detergent.

ACTION OF BILE SALTS ON FATS

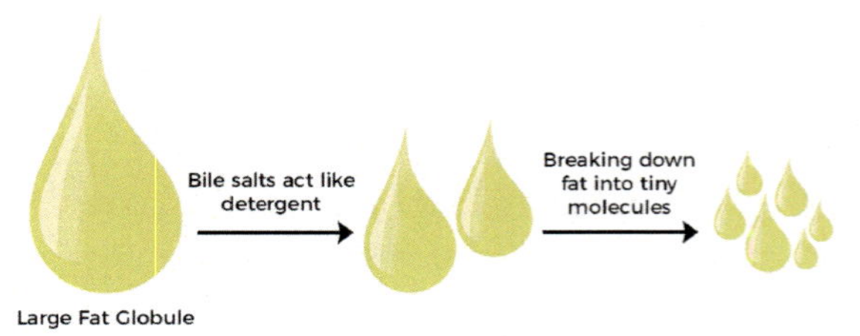

Detergents physically break up grease into small droplets that get mixed into the soapy water and washed away. Similarly, bile mixes with the fats in the small intestine and breaks them up into small fat droplets. These droplets are then broken down and washed away by the digestive enzymes made in the pancreas and washed away into the bloodstream, where they are used as fuel by cells in your body.

Pancreas

The **pancreas** (pan-kree-uhs) is similar to the liver in that it also plays a role in many other bodily processes besides digestion. It looks a bit like a triangle and is found between the stomach and the small intestine.

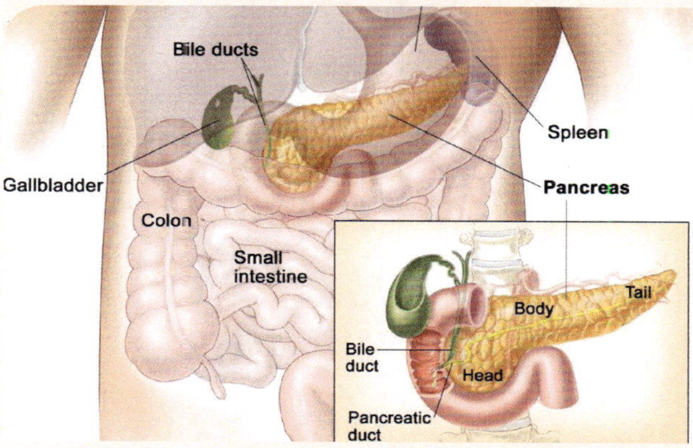

The pancreas is necessary for digestion since it produces the enzymes that flow into the small intestine and finish the breakdown of starches, proteins, and fats. The only thing these enzymes can't break down is fiber, but that's a benefit to us because the fiber actually thickens the nutrient-rich liquid in the intestine and makes it easier for the muscle contractions of peristalsis to push it along. The entire journey through the small intestine takes food between 3 and 5 hours.

Large Intestine

By the end of that long journey, the vast majority of nutrients have been absorbed into the body and what remains of the food you ate moves on to its last destination: the large intestine. The **large intestine** is also called the colon and its primary functions are to absorb some of the water used during digestion back into the bloodstream and prepare the waste materials to exit the body.

ANATOMY OF THE LARGE INTESTINE

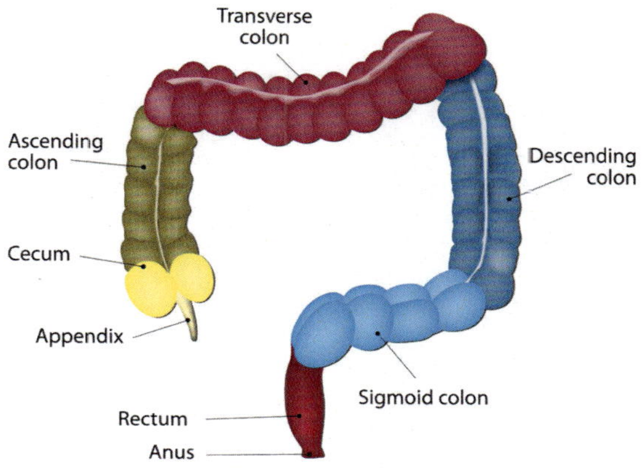

It takes close to 7 liters (2 gallons) of water to digest the food you eat every day. Seeing that most people only drink about 2 liters (0.5 gallon) of water a day, we would be in big trouble if not for the water recycling system Hashem designed to ensure you wouldn't get dehydrated by the digestive process. About 80% of the water secreted with digestive juices gets reabsorbed by the body in the small intestine, as it passes by way of osmosis into the bloodstream along with nutrient molecules. The other 20% of water gets absorbed during the slow passage (about 19 hours long) through the large intestine, which is about 5 feet long (1.5 meters). The water reabsorption is so effective that only 150 milliliters of water is lost per day with the solid waste when it is ultimately eliminated from the body.

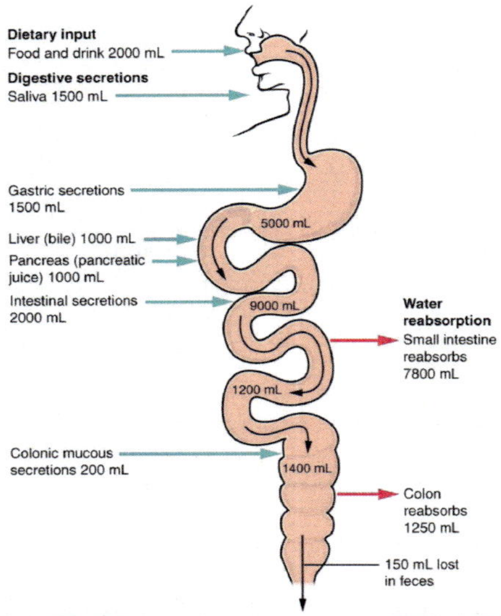

This diagram shows how much water is used during each stage of the digestive process and where that water is reabsorbed into the body.

In addition to its role in recycling water, vitamin K and certain B vitamins are made in the large intestine by an unlikely host of helpers: bacteria! There are trillions of bacteria that call your large intestine home and feed on the materials that pass through it. These bacteria benefit you by producing vitamins that are absorbed into your body along with the water, and help prevent other types of bacteria that could make you sick from moving in.

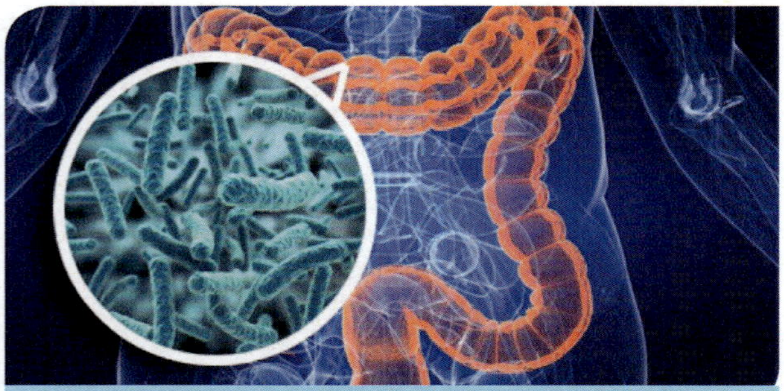

The large intestine is packed with beneficial bacteria.

Once fully processed, waste material moves to the end of the large intestine, called the rectum (rek-tuhm), where it is compressed into a solid form and then eliminated from the body through a muscular opening at its end called the anus (ey-nuhs). The whole process of digestion from start to finish takes about 24 hours! Considering the many miracles you have recently learned about that take place along the way, you can now better appreciate why Chazal set that the conclusion of the blessing of *Asher Yatzar*, said right after the last stage of digestion is completed, should be *"u'mafli la'asos"* (and acts wondrously).

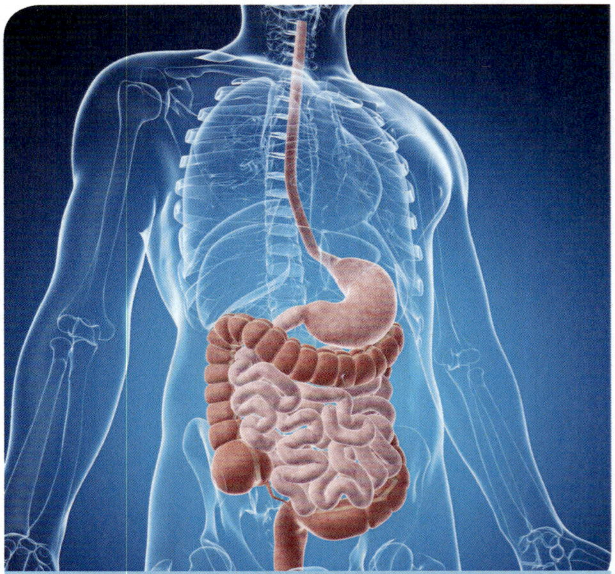

Baruch...asher yatzar es ha'adam bechochmah!

SECTION 1.3 REVIEW

1. What is the role of the small intestine in digestion?

2. What is the name of the yellowish-green fluid produced in the liver that breaks down fat particles? In which organ is it stored?

3. What does the pancreas produce in order to aid digestion?

4. Seeing that the nutrients from food have already been broken down and absorbed by the time they reach the large intestine, how does it aid digestion?

5. What else, besides the waste products of digested foods, is found in the large intestine and why are they there?

USE YOUR MIND

In what ways could digestion be affected by eating too fast?

SECTION 1.4: EXPAND YOUR MIND

SEEING THE PLAN AND THE PURPOSE

There are different opinions among the earlier rabbis as to why Hashem made such a variety of creatures. Rashi explains that when we see so many wondrous works, we are inclined to give praise to Hashem, Who made them (*Rosh HaShanah* 31a). The *Chovos HaLevavos* explains that Hashem made different types of creatures to show us He is capable of doing anything. But after reading our chapter and reflecting upon all of the various types of food and nutrients we need to survive, we can particularly appreciate the opinion of Rabbi Saadiah Gaon (882–942), in his famous philosophical work, *Emunos VeDei'os* (Beliefs and Opinions), who explains that the reason Hashem made everything was for the sake of man—to help obtain his food.

We know that we need carbohydrates, proteins, fats, vitamins, minerals, and water to be healthy—but where do those nutrients come from? Water is the first necessity, which travels by way of the giant, fluffy freight trains we call clouds. Using heat from the sun, salty ocean water is purified. It's moved using the wind and clouds, until it falls as rain and snow onto the ground. After it falls, the water moves through the ground, picking up essential minerals we need. Some of the water is then collected in rivers, lakes, or the natural underground storage facilities for pure water called aqueducts, where it becomes available for us to drink in the form of mineral water.

The majority of rainwater that falls on land (much falls straight back into the ocean) slowly moves its way back to the oceans to start the cycle again. However, some of it is absorbed by the roots of plants along the way, giving them the water and minerals they need to grow.

Once inside those roots, the water combines with air and sunlight to miraculously transform into more leaves or branches. Once grown, the plants are either eaten directly, providing us with the majority of the carbohydrates we need, or are wondrously transformed again into the growing bodies of animals who eat the plants. The animals or their products (dairy foods or eggs) are then eaten, and the same raw materials are transformed again incredibly into the growing bodies of humans. They provide us with the proteins we need to grow, and the fats we need for energy to get around and for insulating our bodies from the cold.

When we think about the food we eat and where it comes from, we can see that Hashem made the sun, clouds, wind, rain, snow, rivers, lakes, soil, rocks, and metals (that contain minerals), and all plants and animals, so that we can have food to eat. So when you make a *berachah* and eat food in order to give you the strength to serve Hashem and do mitzvos, you are literally fulfilling the purpose of all creation!

SECTION 1.4: EXPLORE CHAZAL

TORAS EMES

You probably know that you have to wait six hours between eating meat and milk, but do you know why? This rabbinic law is stated in the Talmud (*Chullin* 105a.) Rashi explains that the reason is because of the fatty nature of meat. Not only does meat leave a residue in one's mouth and throat, but when meat is in the stomach, fatty gasses can escape (through burping, for instance) and cause a "meaty flavor" in one's mouth. If a person would eat dairy during this time, they would have the combined flavor of milk and meat in their mouth, which is prohibited.

With our newfound knowledge of the digestive system, we may be able to better appreciate the astonishing wisdom Chazal had on science, nature, and the workings of the human body. We know that it generally takes a maximum of 6 hours for food to pass through the stomach and into the small intestine. This would explain perfectly why Chazal wanted us to wait 6 hours between meat and milk, as this would ensure no meaty flavors remain in the stomach when starting a dairy meal. By then, almost the only thing left is waste products in the large intestine.

Now, if we look in the Talmud, we find that Mar Ukba comments that he considers himself merely vinegar and his father as wine, since he would "only" wait 6 hours between meat and milk, while his father would wait 24 hours. Although this is not the halachah, we can understand the stringent practice of Mar Ukba's father. Since it usually takes between 24 and 36 hours for food to be fully eliminated from the body after eating, he would not eat dairy until his meat meal had the chance to complete the digestive process.

The more we learn about science and nature, the more we can see the awesome wisdom of Hashem literally everywhere we look, and the more we are amazed at the wisdom of Chazal, which was clearly given to them as a gift. As it says in *Tehillim* (25:14), "*Sod Hashem lirei'av*—The secrets of Hashem are revealed to those who fear Him." It was impossible for them to have known so much based on the limited scientific knowledge of their era alone. But they didn't need to experiment with a hole in someone's stomach in order to view what was going on in there. Hashem gave them a clear picture of the secrets of science and nature, which they needed in order to instruct us how to properly fulfill the Torah.

EXPERIMENT:

LEARNING ABOUT LABELS

Corresponds to Section 1.1

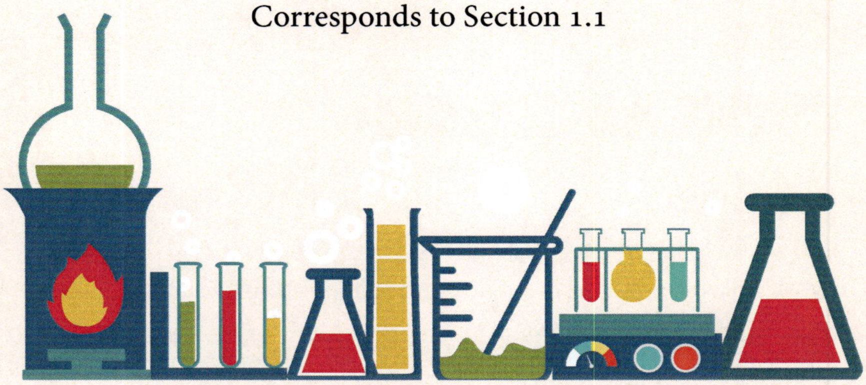

Materials:

Three food labels (can take from any food in your pantry or refrigerator).

Procedure:

Fill out the chart on the facing page to learn more about the different components of a food label.

Questions:

1. If there are 4 calories for gram of carbohydrate, how many of the total calories come from carbohydrates?

(___ grams of carbohydrates x 4 calories per gram = _____ calories from carbohydrates)

2. Which of these three nutrients does your body use first for energy? (Circle one)

A. Carbohydrates B. Fats C. Proteins

3. Which of your two foods seems to be the healthier food?

____

Write food names here			
What food group is this food in?			
What is the serving size for this food?			
Is the serving size realistic? (Is this how much you would normally eat or drink?)			
How many total calories are in one serving?			
How many total grams of fat are in one serving?			
How many grams of carbohydrates are in one serving?			
How many grams of protein are in one serving?			
Name two vitamins and the percentage of the "daily recommended needs" one serving of this food gives you of that vitamin.	Vitamin _______ -- ___ % Vitamin _______ -- ___ %	Vitamin _______ -- ___ % Vitamin _______ -- ___ %	Vitamin _______ -- ___ % Vitamin _______ -- ___ %
Name two minerals and the percentage of the "daily recommended needs" one serving of this food gives you of that mineral.	Mineral: ________-- ___% Mineral: ________ -- __ %	Mineral: ________-- ___% Mineral: ________ -- __ %	Mineral: ________-- ___% Mineral: ________ -- __ %

EXPERIMENT:

ENZYMES IN ACTION AND STOMACH SIMULATION

Corresponds to Section 1.2

Introduction

These experiments are designed to help you better understand the mechanical and chemical digestion that take place in your mouth and stomach. Amylase is a digestive enzyme found in saliva that breaks down starches. In the first experiment, you will examine the action of this enzyme in your own mouth, and in the second experiment you will simulate the churning action that takes place in your stomach.

ENZYMES IN ACTION

Materials

- 3 crackers
- 1 hard-boiled egg (just the white part)
- a clock, stopwatch, or timer to time with

Procedure

1. Put a cracker on your tongue and close your mouth. Leave the cracker there for 10 seconds without chewing it.

2. Take out the cracker, examine it, and then record what happened to it.

3. Put another cracker in your mouth and chew it. In the chart below, record changes in texture or taste that take place, if any, after 30 seconds, 1 minute, and 2 minutes.

4. Put a piece of the hard-boiled egg white in your mouth for 30 seconds without chewing it. Take it out, examine it, and record what changes took place, if any.

5. Put the third cracker in your mouth and chew it for 2 minutes without swallowing. Take it out, examine it, and record what changes took place, if any.

Food	Description
Cracker #1 placed on tongue for 10 seconds	
Cracker #2 after chewing for 30 seconds	
Cracker #2 after chewing for 1 minute	
Cracker #2 after chewing for 2 minutes	
Egg white after sitting in mouth for 30 seconds	
Cracker #3 after chewing for 2 minutes without swallowing	

Questions

1. What happened to the cracker that sat in your mouth for 10 seconds and why?

____

2. As time went on, you should have noticed cracker #2 becoming sweeter in your mouth. Explain why.

____

3. You should have noticed that no changes took place to the egg white while it was simply sitting in your mouth. Why did the cracker begin to dissolve when it came in contact with your saliva, but the egg white did not?

____

4. What did you notice about the taste of cracker #3 after chewing it for 2 minutes?

____

5. What explanation can you think of to explain the change in taste in the cracker after chewing it for 2 minutes?

____

STOMACH SIMULATION

Materials

- 1 slice of bread
- about 8 tablespoons of water
- 2 Ziploc sandwich bags
- Hard-boiled egg (from first experiment)

Procedure

1. Place a piece of bread and 3 to 4 tablespoons of water in the Ziploc bag.

2. Squeeze the bag with your hands repeatedly for 2 minutes, noticing how the water and churning turn the bread into an almost liquid form. This is similar to what takes place in the stomach, as it turns food into chyme before that liquid is passed along to the small intestine.

3. Place the remainder of the hard-boiled egg white into a different Ziploc bag with the same amount of water and repeat the churning.

Questions

1. After mashing, what difference did you notice between the state of the bread and the state of the egg white?

____

2. What substances in the stomach, which were not present in the Ziploc bag, ensure that foods like the egg white get broken down properly and prepared to move into the small intestine?

____

EXPERIMENT:

MODELING THE DIGESTIVE SYSTEM

by Mrs. Sari Ginspar

Corresponds to Section 1.3

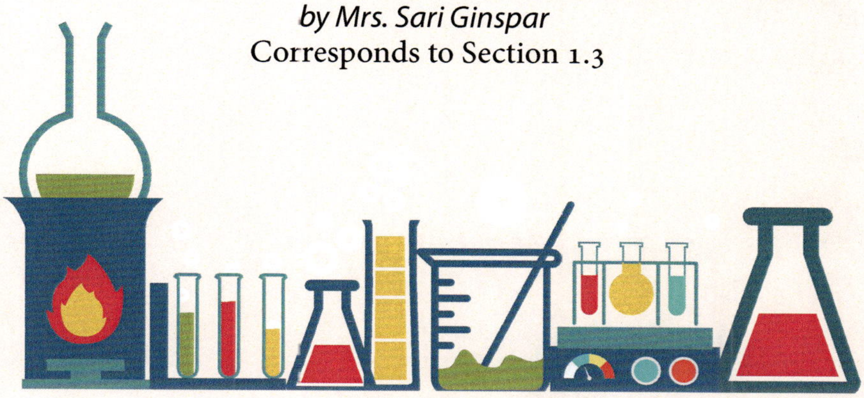

Objectives:

This lab is designed to model many of the process of digestion you have learned about in the chapter. It will generally be to long to be done in one class so should be done over the course of two or more classes.

Background Information:

Teeth Shape and Function: We categorize teeth as one of three types: incisors, canines, and molars. Each kind is shaped differently and has a different job. Incisors are the knife- or scissor-like teeth in the front of the mouth. They act as scissors that cut food into bite-sized pieces so that they can fit into the mouth. Canines are the sharper fang-like teeth that surround our incisors. If you will think about how you bite into food, you will notice that they are used to grip and rip particularly hard or tough foods. Just think of biting into a carrot. We tend to bite into it using our canines. Our back teeth are called molars. They have a large bumpy surface that it perfect for grinding up food.

Materials:

- 1 Small Dish (large enough to use masher in)
- 2 Pieces of Stale Bread (Alternatively, 1 Over ripe banana, cut into 2 pieces)
- 2 Tbsp. of Water
- 1 Scissors or Plastic Knife
- 1 Masher
- 1 Sandwich Size Ziploc Bag
- ¼ cup Vinegar
- 8 Drops of Food Coloring, divided in half
- 1 Knee-High Nylon, or a leg cut off of child's tights, with foot cut off

- 1 Knee-High Nylon, with foot still attached
- 1 Tennis Ball
- 1 Bowl or Small Basin
- 4-5 Sheets of Paper Towel

Procedure:

The Mouth

1. Begin by using the scissors or knife to cut the piece of bread into bite-size pieces and place them into the small dish.

2. Pour in about two tablespoons of water.

3. Use a masher to mash the piece of bread and water.

4. Before answering the questions below, each student should look inside the moth of their partner and take note of the shapes of the different teeth.

Questions:

1. Which kind of teeth are like the scissors or knife used to break the piece of bread into smaller pieces?

   ____

2. What does the water in this model represent?

   ____

3. Which kind of teeth are like the masher? ______________________________

4. What is the difference between the way front teeth and back teeth are shaped?

   ____

5. How is the shape of front teeth suited to their job?

   ____

6. How is the shape of back teeth suited to their job?

   ____

7. How does saliva help with swallowing the food? (Imagine that instead of eating the piece of bread, you are eating matzoh or something else hard and crumbly.)

   ____

8. In this simulation, you moved the food from one place to another with your hands. Which part of your mouth moves the food from your front teeth to your back teeth, mixes the food with saliva and pushes the mixture to the back of your throat to be swallowed?

____

Save the container with the food for "The Stomach" model.

The Esophagus

- Hold the knee-high nylon at one end.
- Place the tennis ball in the top end of the tights.

Questions:

1. How could you get the tennis ball from one end of the knee-high nylon to the other? Does it just fall through the tights or do you have to do something to it?

   ____

2. Suppose the tennis ball represents the food and the tights represent our esophagus. Describe how your body gets the food from the top of your esophagus to the bottom.

   ____

3. What is the purpose of the esophagus?

   ____

The Stomach

- Empty the mixture that you saved into the quart size bag, which represents the stomach.
- Add a ¼ cup vinegar, representing the acid in the stomach.
- Zip the bag shut and use your hands to mix the food and vinegar in the bag.
- Add 4 drops of food coloring. In our model, this represents pepsin, an enzyme that breaks down protein. The color change of the food represents the pepsin breaking the bonds in the food.

Questions:

1. Open the bag and smell the contents. Describe the smell.

   ____

2. What does the mushing of your hands represent?

   ____

Add 4 drops of food coloring to a fresh piece of bread.

Questions:

1. Compare the color change from the whole piece of bread to the color change from the mushed piece of bread.

   ____

2. Why is mechanical digestions so important for chemical digestion to occur properly?

   ____

The Small Intestine

- Place a fresh knee high nylon, with foot still attached into the basin. Pour the mixture into the nylon. Liquid will leak out through the material. Squeeze the nylon slightly.
- The basin that surrounds the nylon represents blood vessels that will bring the digested food to all of the parts of the body. The nylon represents the small intestine.

Questions:

1. Does everything leak out of the nylon into the basin?

   ____

2. What determines whether the food leaves the nylon or stays inside?

   ____

3. Why does food need to be digested into such small pieces?

   ____

The Large Intestine

- Although the food moves from the small intestine into the large intestine, we will continue to use the same nylon. Now the nylon will represent the large intestine.
- Lay the nylon on multiple paper towels layers and roll it up tightly, squeezing and patting as you go.

Questions:

1. Is the material left in the nylon at this point food or waste? How do you know?

   ____

2. The material left in the "large intestine" is being squeezed dry. What could happen if it weren't squeezed dry?

   ____

Questions about Modeling:

1. Name three concepts about the digestive system that were modeled in this lab.

   ____

2. While making models helps us understand scientific processes, models have their limitations. After you learn more about the digestive system in text, list three concepts that were not included in our model.

   ____

   ____

   ____

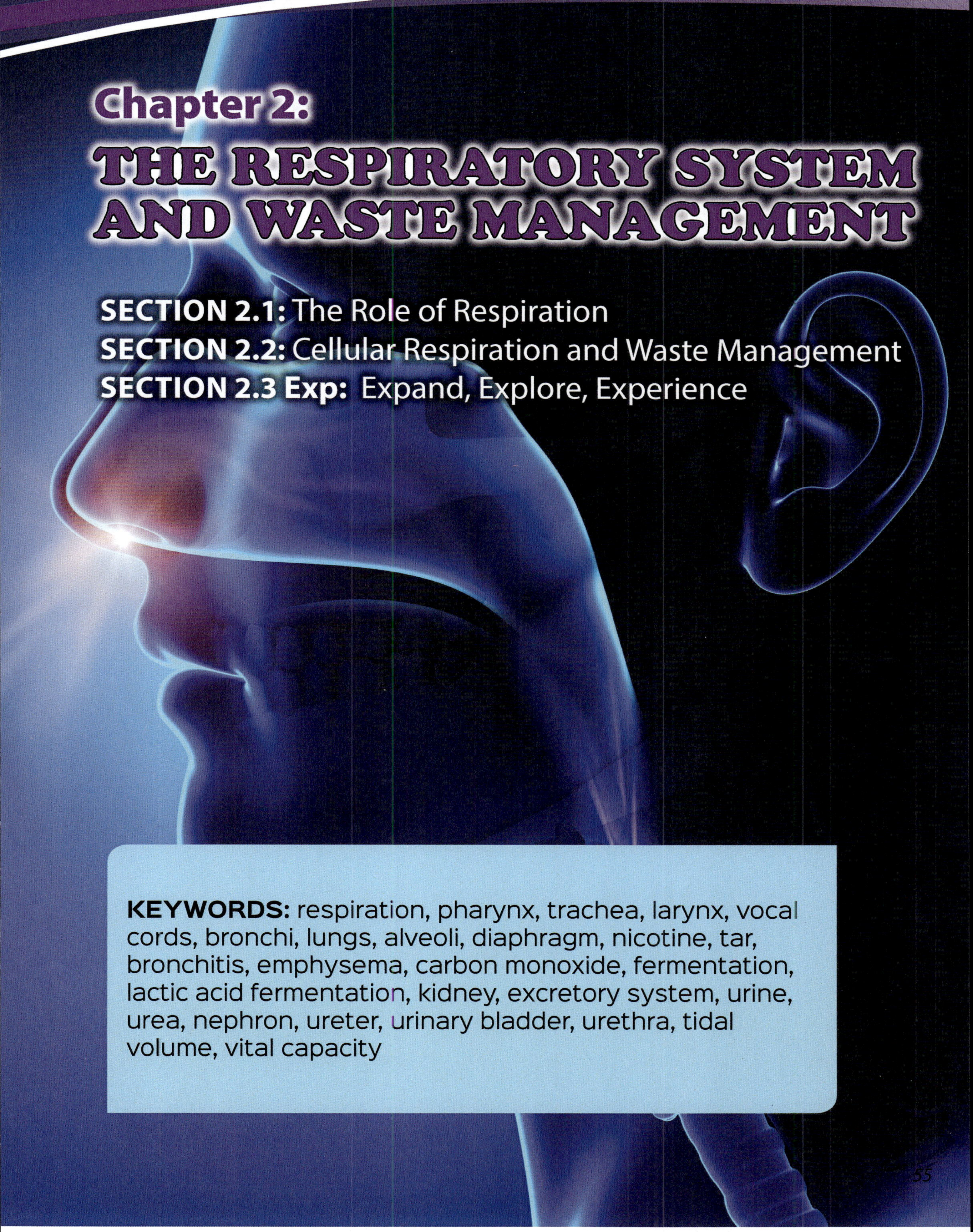

Chapter 2:

THE RESPIRATORY SYSTEM AND WASTE MANAGEMENT

SECTION 2.1: The Role of Respiration
SECTION 2.2: Cellular Respiration and Waste Management
SECTION 2.3 Exp: Expand, Explore, Experience

KEYWORDS: respiration, pharynx, trachea, larynx, vocal cords, bronchi, lungs, alveoli, diaphragm, nicotine, tar, bronchitis, emphysema, carbon monoxide, fermentation, lactic acid fermentation, kidney, excretory system, urine, urea, nephron, ureter, urinary bladder, urethra, tidal volume, vital capacity

SECTION 2.1: THE ROLE OF RESPIRATION

Now that you know how food gets eaten, digested, and absorbed into the body to give you energy and turn into little pieces of you, you are ready to explore how oxygen, the most vital and important element of all, is obtained.

Astronauts need to wear special suits with oxygen tanks so they can explore outer space.

Oxygen is more important than food, and even water, since a person can survive days without either one, while a person cannot live without oxygen for more than a few minutes. The reason why oxygen is so important is that the energy-releasing chemical reactions that must occur constantly in the cells of your body will not take place if oxygen is absent. Cutting off oxygen to the body is similar to turning off the engine of a car. Inside your body, the primary source of fuel comes from the broken-down components of carbohydrates, which are single glucose sugar molecules. While the body may still have fuel (glucose), without the ability to burn the fuel, it cannot move.

The process through which sugar molecules (glucose), in the presence of oxygen, undergo a complex series of chemical reaction inside your cells to give you energy is called **respiration** (res-puh-rey-shuhn). The term "respiration" also refers to the act of breathing or the movement of air into and out of the lungs. In order to differentiate between the two processes, the respiration that occurs inside your cells is often referred to as chemical respiration. For the sake of clarity, we will refer to breathing as respiration and the process of burning the fuel obtained from food inside your cells as chemical respiration.

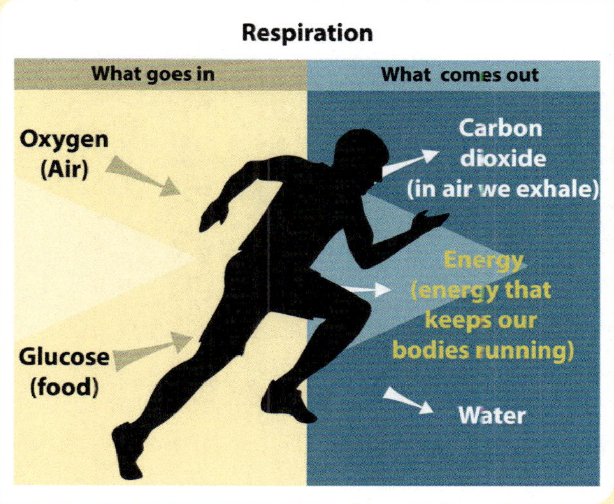

There are a number of organs that work together to carry out the process of respiration; they are collectively referred to as the respiratory system. The respiratory system has two main functions. The first is to bring oxygen from the air outside into the body, where it can be distributed to the cells via the bloodstream. The second function is to remove the toxic by-product of chemical respiration, which is carbon dioxide. While carbon dioxide can be viewed as a waste product of chemical respiration, it by no means goes to waste. Trees and plants take in the carbon dioxide we exhale and use it, along with sunlight and water, to make sugars and food that both they (the plants) and we get to eat. The fact that the waste products of humans are needed by plants in order to produce things that benefit both the animal and plant kingdoms is testimony to the ingenious design of our world. It's also something fun to think about and appreciate every time you breathe!

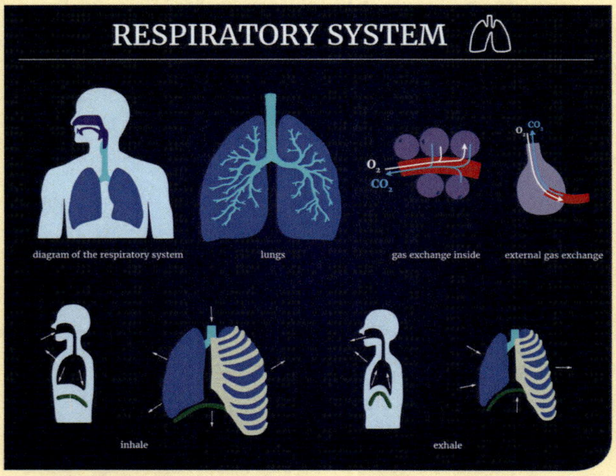

Following a Breath

Respiration begins with breathing. It is not something you have to think about doing and is automatically controlled by your brain to ensure you get the 20,000 breaths you need every day to get the oxygen you require. There are a number of organs involved in the breathing process, facilitating breathing in air (inhalation) and breathing it out (exhalation), as well as the gas exchange of oxygen and carbon dioxide that take place during each breath. Let's follow the path of a single breath to find out how this amazing process is accomplished, and get to know the organs involved a bit better.

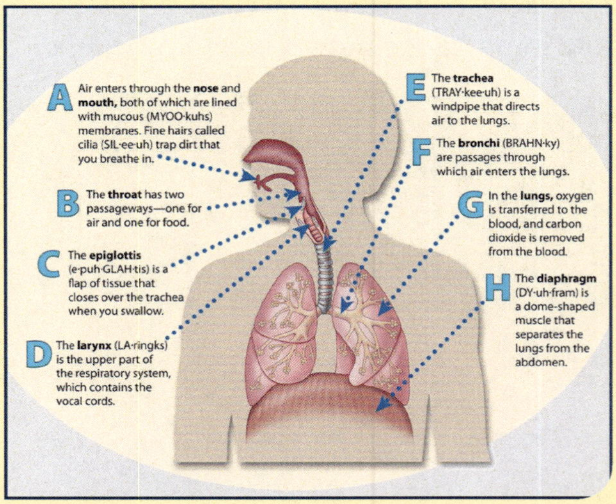

Although air usually looks quite clear and clean, there are, in fact, many particles other than oxygen floating around that you don't want or need. If you have ever noticed the rays from the sun shining in through a window and saw a host of dancing particles in the air, you have seen some of the unwanted elements in the air. These elements include grains of dust, pollen from plants, and ash from fires.

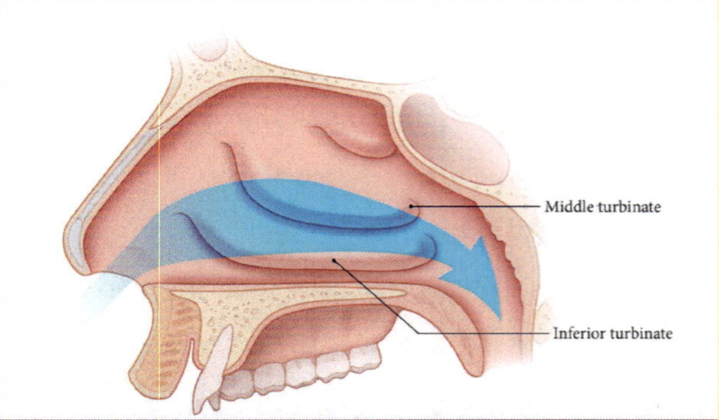

Bony elements called turbinates form the upper chambers of nasal cavities and increase their surface areas. The extra space enables more air to come in contact with the nasal cavities, as well as rapid warming and moistening of air as it passes to the lungs.

Air purification begins with the hairs inside your nose. The hairs were designed to catch large particles coming in with the air and act like a filter to keep out dust, dirt, and germs that could make you sick.

Once inside, air moves into spaces called nasal cavities that appear grooved due to a series of curved bones called turbinates. The warm, moist environment of the nasal cavities warms and moistens the air as it enters to prevent cold or dry air from entering the lungs and damage the delicate structures found there. The nasal cavities also act to purify the air due to specialized cells that produce mucus. Foreign particles coming in with the air get caught in the sticky goo.

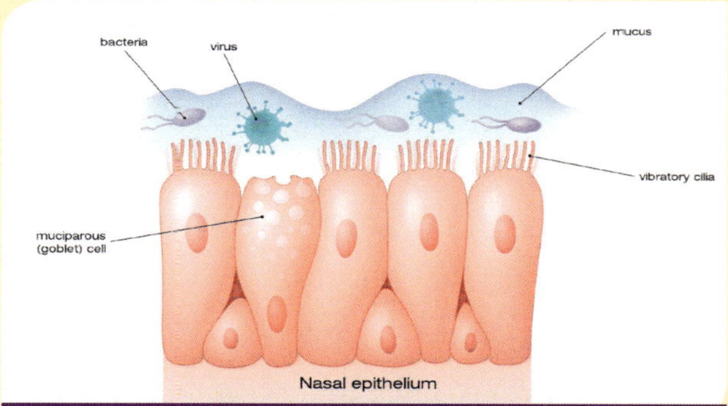

Diagram of the cells lining the nasal cavities

To dispose of the waste, the nasal cavities have other cells with hair-like extensions called cilia that move together in a sweeping motion. The cilia continuously sweep the mucus, along with the foreign particles collected, into the throat. When you swallow, the contents go down into your stomach, where the mucus and everything trapped inside it is destroyed by the hydrochloric acid there. When there is noticeable buildup of mucus that comes from illness or any other reason, the mucus generally contains bacteria or other foreign particles. This should not be swallowed, but blown out into a tissue.

After the nasal cavities, air enters the **pharynx** (far-ingks), or throat. Here it is further cleaned, as particles stick to its moist walls and are swept away by cilia, moistened and heated. The pharynx is actually the only part of the respiratory system that is shared with the digestive system, since food also passes through the pharynx on its way to the esophagus and stomach. Once through, air moves into the **trachea** (trey-kee-uh), or windpipe.

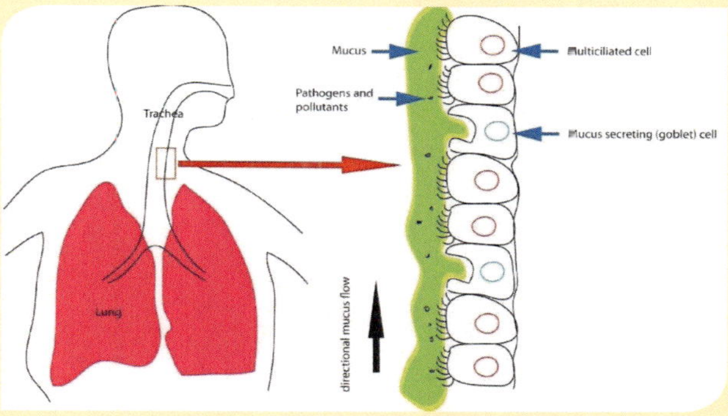

The trachea is a tube with a series of ridges made out of a type of tough elastic tissue called cartilage (more on that in chapter 4). Cartilage is firm enough to keep the trachea open so air can move through it, yet elastic enough that you can turn your head and bend your neck. You can feel your trachea if you want to by running your fingers gently down the center of your neck. It is similar to the nasal cavities of the nose in that it is lined with mucus and cilia that continue cleaning and moistening the air. All of the cilia in the trachea sweep upward and carry mucus laden with dust and germs back up to the pharynx, where it is swallowed.

At the top of the pharynx lies a very important organ that enables you to speak. It's called the **larynx** (lar-ingks), or voice box. The way it works is similar to how guitar strings make music. When someone plays a guitar, the strings vibrate, which causes the air molecules around them to vibrate and produce sounds. You have two strings in your larynx called the **vocal cords**. They are made out of connective tissue that stretches across the opening of the larynx and they enable you to speak.

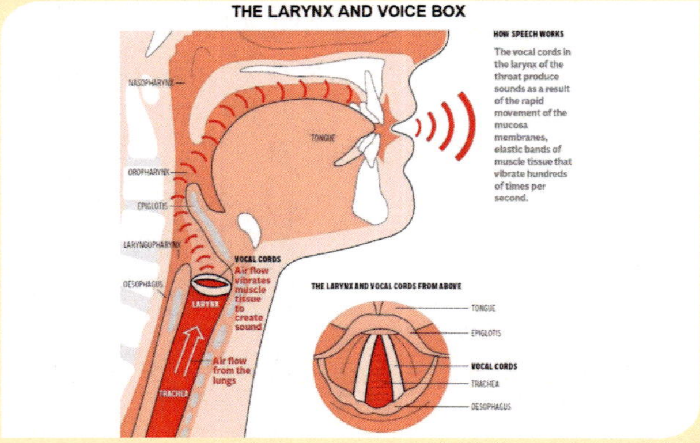

When you want to talk, muscles attached to the vocal cords contract and narrow the opening to the larynx. As air coming from the lungs rushes through this narrow opening, it makes the vocal cords vibrate and produce the sounds of your voice. The length and width of a person's vocal cords affect the sound of their voice. For instance, girls' voices are generally higher pitched than boys' because girls' vocal cords are usually shorter and thinner than boys'. Similar to how shorter strings of a guitar will produce higher-pitched sounds, the shorter vocal cords found in girls produce higher-pitched voices.

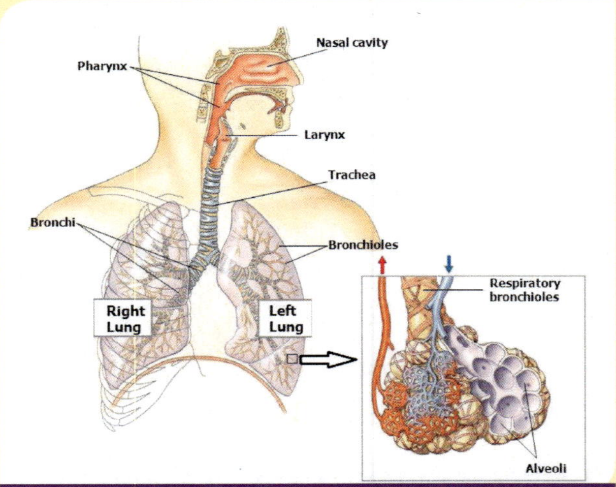

The smallest branches of the bronchi are called bronchioles. That is where gas exchange takes place inside alveoli.

After the air passes through the trachea, it branches out into two passages called **bronchi** (brong-kee; singular: bronchus). The bronchus on the left side of the body goes to the left lung and the one on the right side to the right lung. Each bronchus then divides into many smaller and smaller branches called bronchioles that, if stretched out, would measure about 1,500 miles (2,400 kilometers). These airways were designed to be so numerous so as to maximize the amount of air they can hold and gas exchange they can accomplish. The bronchi then branch out into the **lungs** (luhngs). These are the main organs of respiration and the place where the oxygen from the air comes into the blood and carbon dioxide goes out.

The action of gas exchange between oxygen and carbon dioxide takes place in small sacs called **alveoli** (al-vee-uh-lahy; singular: alveolus). These sacs look like clusters of grapes and are made out of lung tissue designed to move gases between the air and the blood. Because the walls of the alveoli and the microscopic blood vessels called capillaries that surround them are so thin, very small molecules like oxygen, carbon dioxide, and water can pass into and out of the alveoli from the blood.

GAS EXCHANGE IN HUMANS

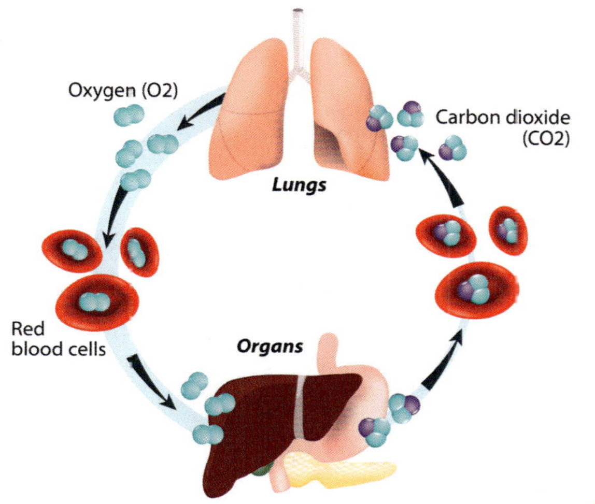

Since the air coming into the alveoli has a higher concentration of oxygen than that in the body, because of all the oxygen used up by cells, it passes into the blood by way of diffusion. Diffusion is a natural process where substances flow from a higher concentration to a lower concentration. Similarly, the blood coming out from the heart into the lungs has a higher concentration of carbon dioxide, as it picked up much while flowing through the body, and therefore diffuses out of the blood into the air sacs of the alveoli. In short, with each breath you take, oxygen is moving into your blood and carbon dioxide is being sent out.

ALVEOLUS GAS EXCHANGE

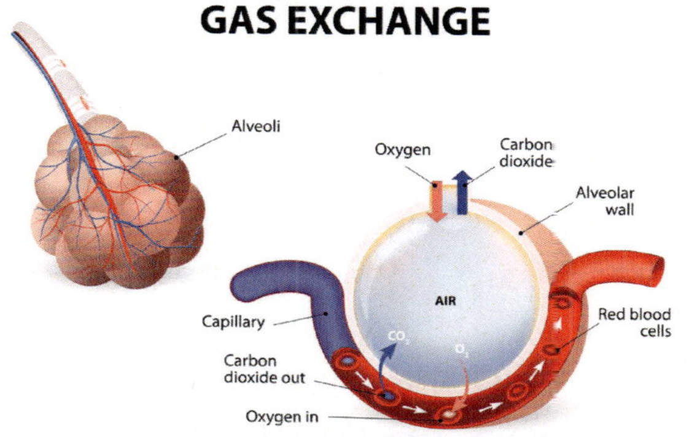

The 300 million alveoli in each lung create a surface area of about 750 square feet (70 square meters), enabling you to take in a lot of oxygen with each breath.

How You Breathe

The organs that get the air into your lungs, so that gas exchange can take place, are clearly vital to respiration. But there are other parts of your body that participate in physically taking oxygen-rich air in and pushing the air carrying carbon dioxide out. The main organ that accomplishes this task is a large, dome-shaped muscle found right underneath your lungs called the **diaphragm** (dahy-uh-fram). The diaphragm works along with the muscles around your ribs to expand and contract your chest, causing air to move in and out with each breath.

The diaphragm functions in breathing

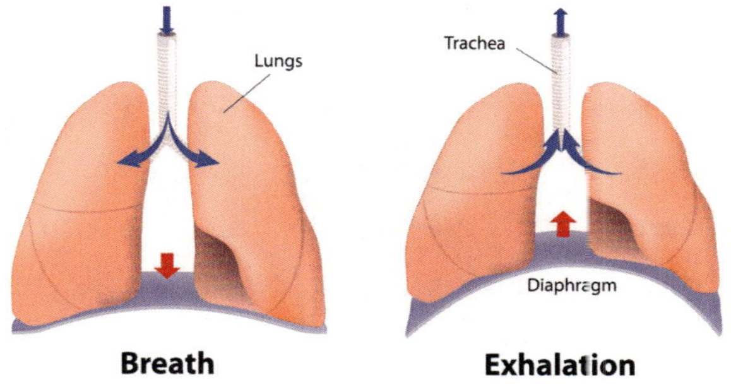

When you inhale, the diaphragm contracts, causing it to move downward. At the same time, the rib muscles also contract, which lifts the walls of the chest upward and outward. The combined action of these muscles causes the chest cavity to expand. Now that the chest cavity is larger, the air that was there before the expansion now occupies a larger space, which decreases the air pressure (density of the air) within the chest. How does that affect us? You'll find out soon…

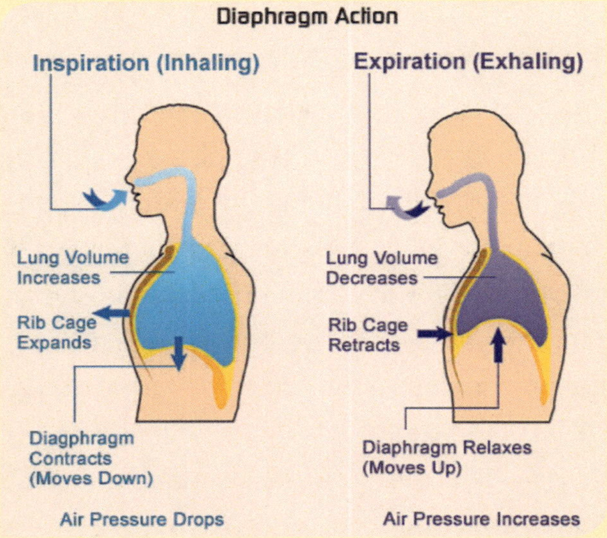

Air might seem empty and weightless, but it is actually full of different types of atoms, like oxygen and nitrogen, that cause it to weigh down or push on anything it is sitting on or inside of. When you breathe in and your chest cavity expands, the air pressure inside your chest becomes lower than the pressure in the air outside of your body. Whenever there is a difference in pressure, air always travels from an area of high pressure to that of low pressure, so inhaling causes air to rush inside your chest.

The exact opposite happens when you exhale, or breathe out. The rib muscles and diaphragm relax, which reduces the size of the chest cavity. This literally squeezes the air out of your chest, as the air pressure inside of you becomes higher than the air pressure outside.

It is pretty amazing, if you think about it. Every few seconds air moves in, gets warmed, cleaned, and moistened; goes into your lungs, where oxygen from the air moves into your blood and carbon dioxide from your blood is moved out to your lungs; and then finally pushed out as you exhale. You probably had no idea there was so much going on literally every time you breathe!

Sneezing, Coughing, and the Hiccups

By now, you know way more than most people about your respiratory system, but there are still a few things that happen to you regularly that you would probably like to know more about, like sneezing, coughing, and the hiccups. Sneezing and coughing are both abilities Hashem gave your body to get rid of dirt, dust, and germs.

A sneeze in action

A sneeze is generated when bacteria or other foreign particles land on and irritate the cells lining your nose or throat. This triggers the powerful force of a sneeze to send the unwanted guests flying. Coughing is also caused by bacteria or other particles that get into the lungs or trachea. While it—*cough, cough, cough*—may be annoying, coughing enables you to get germy mucus out of your lungs when you are sick.

Hiccups, on the other hand, are quite mysterious. No one really knows what benefits they bring, although their cause has been identified. When the diaphragm gets irritated, for example after a person eats too much or too quickly, feels nervous or excited, or drinks carbonated beverages, it can contract uncontrollably. This forces you to suddenly suck air into your throat, where it hits your voice box. The rapid rush of air makes your vocal cords close suddenly, creating the distinct "*hic!*" sound.

Smoking and Your Health

While respiration is a superb system of bringing fresh air full of oxygen into your lungs and sending carbon dioxide out to the trees, it is also quite delicate and can easily become damaged by a very bad habit practiced by many people around the world: smoking cigarettes. The main ingredient in cigarettes are the dried-out leaves of the tobacco plant. Tobacco contains a substance called **nicotine** that is a type of drug called a stimulant. Nicotine makes smokers feel good by stimulating the nervous system, the heart, and other organs. Nicotine is addictive, meaning that smokers develop a physical dependence on it, and have intense cravings for cigarettes if they ever try to stop. Even after "only" smoking cigarettes for a short time, smokers begin to pay a potentially deadly price for their pleasure. Even smoking one cigarette can be dangerous, as people think they'll just try it "once" and it often leads on a path to nicotine addiction.

Due to its dangers, cigarette companies are required to put warnings on every package, stating the effects smoking can have.

Tobacco contains a number of deadly chemicals, such as tar and carbon monoxide, that can lead to numerous health problems and diseases. Smoking is so dangerous that every year in the United States, more than 400,000 people die from smoking-related illnesses. Furthermore, in 2008, the World Health Organization named tobacco as the world's single greatest preventable cause of death, due to the over 6 million people that die from smoking each year. That's 16,000 smoking deaths every day!

Picture 16, Caption: The buildup of tar actually makes the lungs of smokers turn black, and often causes diseases.

One reason why smoking is such a killer is due to the dark, sticky substance called **tar** that forms when tobacco burns. When someone inhales cigarette smoke, the tar sticks to the cilia that line the trachea, bronchi, and smaller airways leading to the alveoli. When coated in tar, cilia cease to move and can no longer sweep out harmful materials or prevent them from getting into the lungs. This often leads to the ongoing or frequent occurrence of the illness **bronchitis** (brong-kahy-tis), called chronic bronchitis, where the small air passages in the lungs get clogged with mucus. That makes it difficult to breathe and can cause permanent damage to breathing passages.

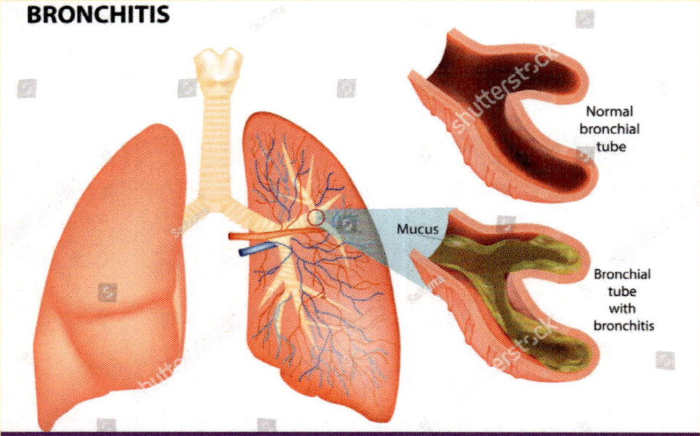

Internal effects of chronic bronchitis and emphysema

In addition to harming breathing passages, chemicals in tobacco can also damage lung tissue and result in a disease called **emphysema** (em-fuh-see-muh). Due to their damaged lung tissue, those suffering from emphysema do not get enough oxygen and cannot get rid of carbon dioxide adequately when they breathe—which makes them continuously short of breath. Once a person develops emphysema, the lung tissue can never be repaired, even if the person manages to quit smoking. Cigarette smoke is so deadly that even being exposed to secondhand smoke (breathing in the smoke exhaled by smokers) for prolonged periods can cause permanent lung damage.

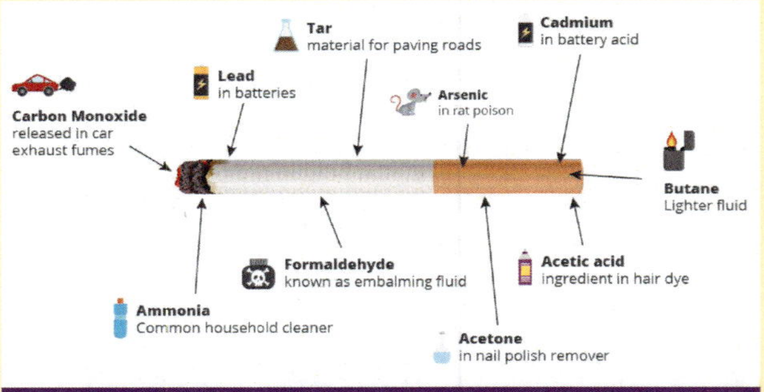

There are over 250 harmful chemicals in cigarette smoke. Here are a few of them and other places they are found.

Another deadly substance that is produced when tobacco is burned is **carbon monoxide**, a poisonous, odorless, and colorless gas. Carbon monoxide binds to the proteins inside red blood cells that are designed to carry oxygen to the cells throughout the body. When the red blood cells, which contain a protein called hemoglobin, are full of carbon monoxide, the blood of smokers has less room to carry oxygen. Their hearts are therefore forced to work harder and beat faster to get enough oxygen to meet their needs. The increased levels of carbon monoxide in smokers' blood makes it difficult for them to perform daily activities that require effort, like climbing stairs, since they get short of breath quickly. It also increases their chances of developing a heart disease such as heart attacks and atherosclerosis.

In addition to all of the above, there are numerous cancer-causing agents in cigarette smoke. About 140,000 Americans die each year from lung cancer, and the vast majority of those cases are caused by smoking. Many of those who do not die from lung cancer suffer horribly from cancerous growths called tumors that grow in the lungs and take up space that would otherwise be used for gas exchange. Lung cancer is particularly dangerous, as it is difficult to detect and is most often found when it is too late to treat effectively.

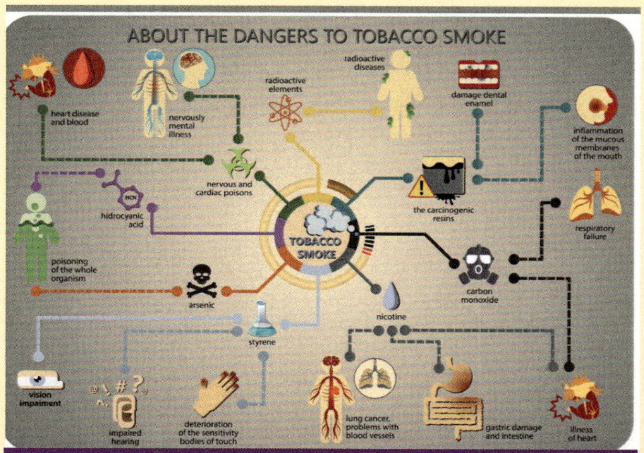

Summary of health issues caused by smoking tobacco

So, if someone ever offers you a cigarette or you are in a place where it seems normal to smoke because many other people are doing it, remember the serious health problems those smokers are bringing on themselves. Be wise and have mercy on the delicate and precious lungs Hashem gave you by saying no to tobacco.

SECTION 2.1 REVIEW

1. Which two different processes does respiration refer to?

____

2. What happens to air as it travels through your nose and nasal cavities?

____

3. What process takes place in the lungs?

____

4. Which muscles are involved with breathing, and how do they accomplish this?

____

5. What are some of the health risks of smoking tobacco?

____

USE YOUR MIND

What type of response could you give to a friend who wants you to try smoking a cigarette?

____

____

SECTION 2.2: CELLULAR RESPIRATION AND WASTE MANAGEMENT

Your oxygen-filled blood always travels from the lungs to your heart first. From there, it gets pumped throughout your body, supplying all of your cells with the oxygen they need. When oxygen enters your cells, it is used by the mitochondria, the cellular organelles that are the site of energy production, to enable a complicated series of reactions that convert the energy in sugar molecules (glucose) into ATP (adenosine triphosphate)—a form of fuel that runs your microscopic cellular machinery.

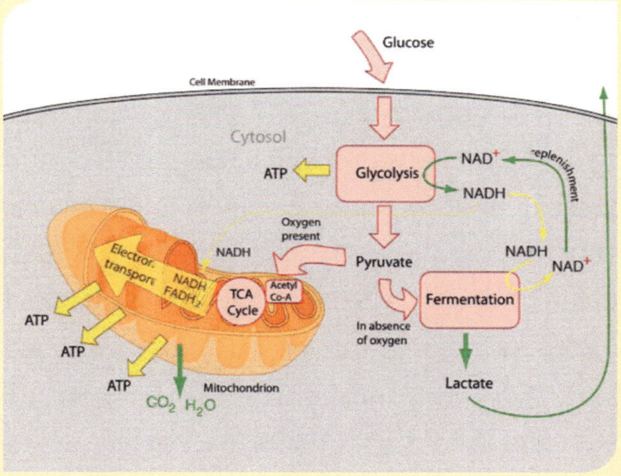

The process of cellular respiration has two stages, one that happens in the cytoplasm (all the material inside a cell other than the nucleus and organelles), and the other in the mitochondria. During the first stage, glucose molecules are broken down into smaller molecules. Oxygen is not needed for this to occur and there is not much energy released (only 2 ATPs). However, for the second stage to occur, oxygen must be present. The second stage takes place in the mitochondria, where the molecules are broken down further and massive amounts of energy that the cell can use are produced (36 ATPs).

During stage one, when oxygen is not present, the cell can still make a little bit of ATP, but not nearly enough to supply the body with all of its needs. But even this modest amount can be very helpful when your body is being pushed to the max and you need extra energy to keep going, such as during intense exercise. The release of energy from food without the use of oxygen is called **fermentation**. When it happens in people, it is referred to as **lactic acid fermentation**, due to the lactic acid that is produced as a by-product of the energy-releasing process.

The painful or burning sensation you feel in your muscles when exercising for extended periods of time is actually the buildup of lactic acid produced from fermentation. It is also a sign that your cells were using more oxygen than can be taken in no matter how fast you breathed. The ability to carry out lactic acid fermentation was designed to allow people to keep on going when circumstances call for it.

When you "feel the burn," it is a sign that you are getting a great workout, although many recommend stopping the exercise if you feel real pain and not just minor discomfort.

Waste Management

In the presence of adequate food, water, and air, the body is able to carry out all of its processes smoothly, performing the trillions of daily tasks that are required for life. With all of the "cellular cities" running 24 hours a day, there are numerous waste products produced. As we already discussed, one of them is carbon dioxide, which is a by-product of cellular respiration and is removed from the body by way of the lungs when you exhale. However, the majority of waste products produced by cells, such as lactic acid, require a different waste management system to take superb care of taking out the trash.

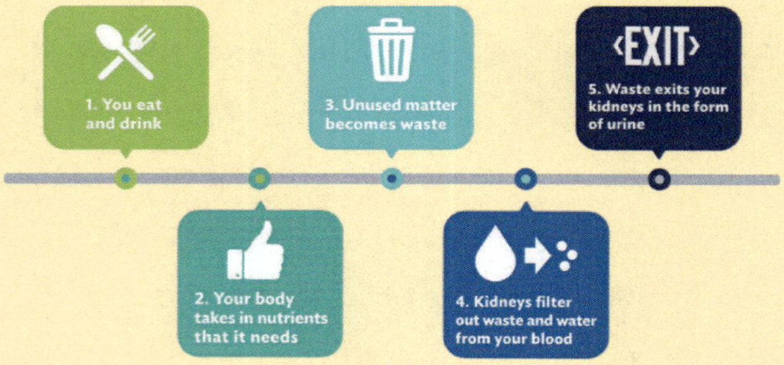

Hashem, of course, did it again by giving us two incredible filtration and waste removal organs called the **kidneys** (kid-neeys). The kidneys are the main organ in what we call the **excretory system**, which that is in charge of collecting waste products produced by cells and removing those wastes from the body along with excess water. The kidneys carry out this great task by working together with accessory organs that you will learn about shortly: the ureters, urinary bladder, and urethra.

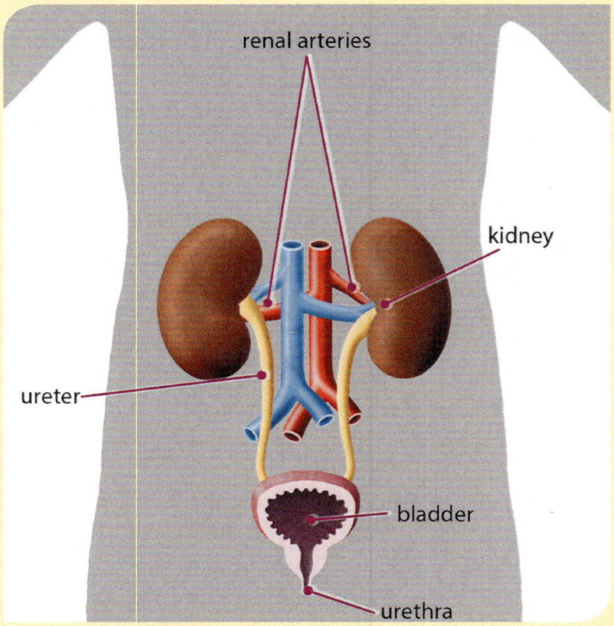

All of the blood in your body (which totals about 1.3 gallons or 5 liters) can pass through your kidneys in under an hour, and blood is constantly passing through them 24/7. As it passes through, the blood is carefully checked in the kidneys. Waste products are removed and eliminated through the **urine** (yoor-in), a watery fluid that contains excess water and waste products. Although there are many waste products produced by cells, the most common one is called **urea** (yoo-ree-uh), a chemical produced from the breakdown of proteins.

Filtration

You have two kidneys that work together to filter your blood. Each of them could actually accomplish the job on its own, which is why some kind and caring people are able to donate one of their kidneys to save the life of someone else whose kidneys no longer function. Each kidney accomplishes the fantastic feat of filtration by way of a million or so microscopic filtering factories called **nephrons** (nef-rons). Nephrons are designed to regulate the chemical composition of the blood. They accomplish this by way of a two-stage filtration system, which enables the kidneys to monitor the level of water and wastes present in the body. It may take a bit of focus, rereading, and review, but it's worth understanding how Hashem keeps your internal body clean. Let's explore this process in depth, by following the flow of blood as it enters the kidneys.

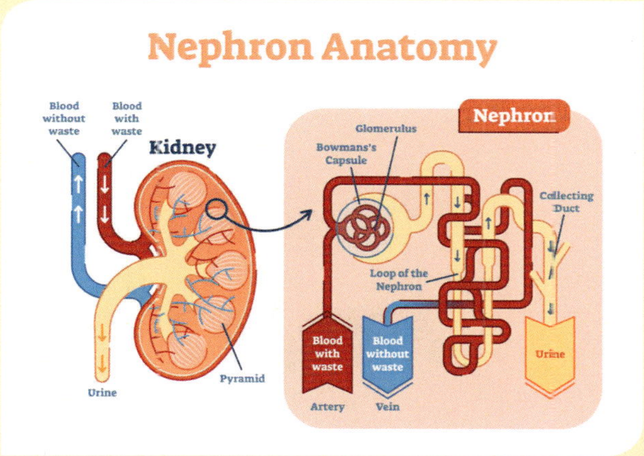

During the first stage of filtration, waste products along with some materials that are needed, such as sugars, are filtered out of the blood. This happens when blood coming into the kidneys flows through smaller and smaller blood vessels until it reaches a cluster of capillary vessels that are only one cell thick, called the glomerulus (gloh-mer-yuh-luhs). Because the capillaries are so thin, larger molecules like proteins and red blood cells cannot pass through their walls and therefore remain in the blood. Smaller molecules like water, urea, and glucose pass through them into a capsule surrounding the capillary cluster called the Bowman's capsule.

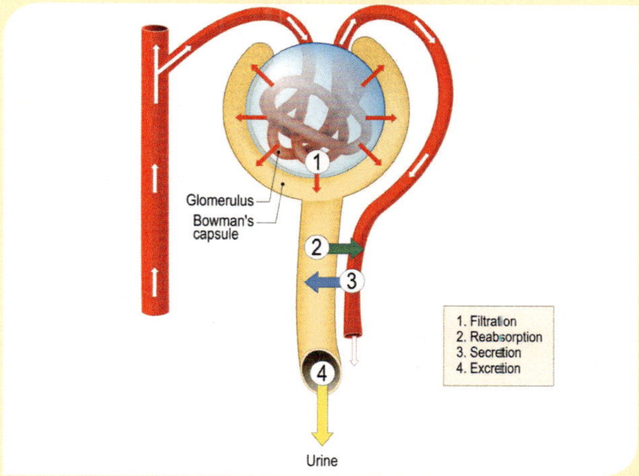

The capsule where the waste products, water, and other components are collected is connected to a long, twisting tube. During the second stage of filtration, the liquid moves through this tube and many of the substances that were originally filtered out are returned to the blood. Under normal conditions, all of the glucose, most of the water, and small amounts of other materials pass back into the blood by way of capillaries surrounding the tube. In contrast, urea and other waste products remain in the tube and flow out of the kidneys as urine passes into two narrow tubes called **ureters** (yoo-re-ters).

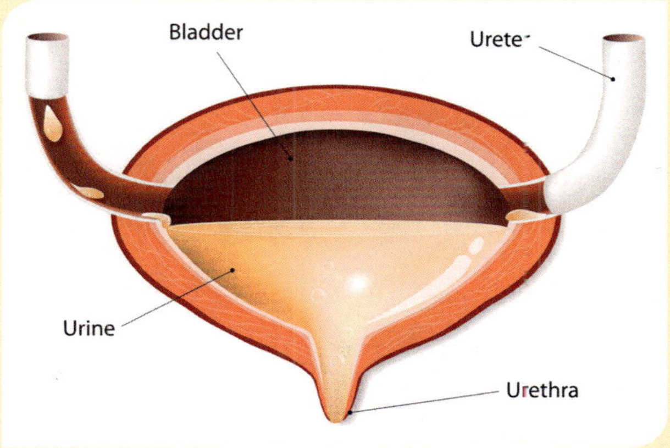

The ureters carry urine to the **urinary bladder**, which is a saclike muscular storage organ. Urine eventually leaves the body through a small tube called the **urethra** (yoo-ree-thruh).

That happens after our built-in sensors which are part of the nervous system alert us when it is full by giving us the feeling that we need to use the restroom. The fact that this entire process takes place without thought—yet the storage and release of the contents in the urinary bladder is under your control—is a great kindness from Hashem. It affords people the dignity to find a private place to relieve themselves. Those who are not well and have lost conscious control of this process know just how great of a gift this is. They would pay any price to be able to pass waste at a time and place of their choosing. We, who do have the precious and dignified gift of bladder control, should take a little bit of time to say thank You to the One Who keeps our intricate purification and excretion systems running smoothly.

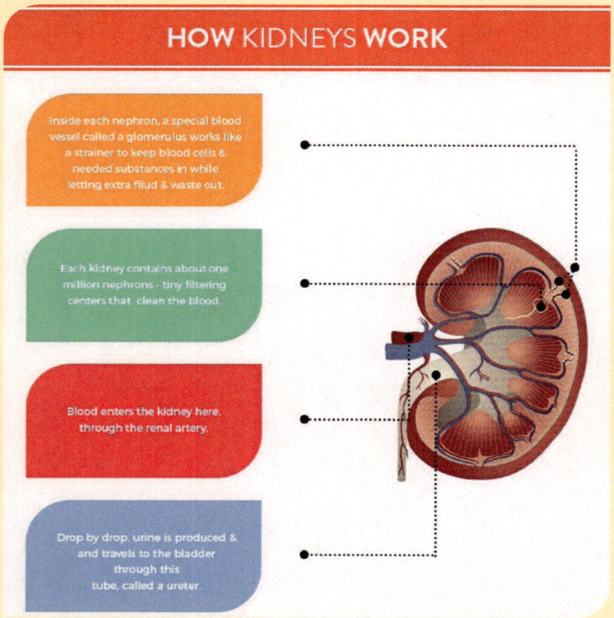

As we mentioned above, the two-staged filtration system of the kidneys enables regulation of the water levels inside the body in addition to filtering out wastes. It is pretty obvious that you don't want hazardous substances in your bloodstream for long, so the kidneys ensure that does not happen. Too little water is also a commonly known danger called dehydration—that's when there is inadequate water available to supply cells with their needs. Dehydration causes cells to shrivel and not work as well. In extreme cases, dehydration can cause fainting and even death. When water levels are low, such as on a fast day, the kidneys do their best to prevent dehydration, by sending much, if not most, of the water back into the bloodstream to conserve as much water as possible.

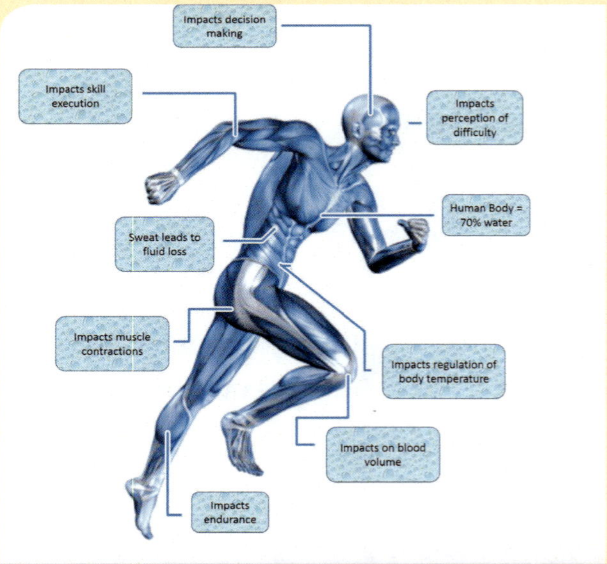

Some of the reasons why it is important to have adequate water in the body

What is a little less understood is how having too much water in the blood can be dangerous as well. When too much water is present, it can cause excessive diffusion into cells, causing them to burst.

One of the most important roles of the kidneys is to make sure this doesn't happen—by sending out excess water as urine. Doctors recommend drinking at least 2 liters of water per day for those over the age of 13. That quantity gives the cells all the water they need to work properly. The kidneys make sure that the optimal amount of that water gets into the blood. They also make sure that many other substances, such as sodium, potassium, and calcium, are kept at the right levels in the blood. The removal of waste products and monitoring of water and other substances in the blood by the kidneys is one of many regulatory mechanisms Hashem created in order to maintain the ideal internal conditions of the body. (This state, known as homeostasis, will be discussed more in chapter 5.)

Excretory Organs of the body

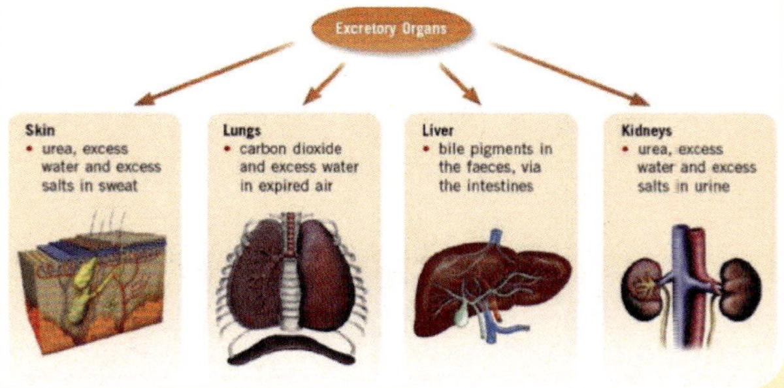

In addition to the kidneys and lungs, there are two other organs that play a role in excretion: the liver and the skin. The liver plays an active role in excretion by forming urea. Although urea itself needs to be excreted, the original waste product from the breakdown of proteins is ammonia, more commonly known as bleach. Due to the great danger of having bleach running through your bones, the liver quickly transforms it into a more manageable form, urea, which is then removed from the body by way of the kidneys. The liver also removes substances such as cholesterol and waste from medicines by converting them into forms of bile which are sent first to the gall bladder, then to the small intestines and finally removed with the rest of the waste products from foods.

The skin is also considered part of the excretory system, since both urea and water are excreted when you sweat. It is for this reason that halachah requires one to keep safe by never ingesting human sweat—or even putting food inside one's clothes, where it could come in contact with the skin, absorb some of the toxic materials found in sweat, and be ingested (*Kitzur Shulchan Aruch* 33:3).

SECTION 2.2 REVIEW

1. When oxygen is not present, how do the cells in your body produce energy?

____

2. What is the function of the kidneys?

____

3. Which other organs work together with the kidneys?

____

4. What is a nephron and what does it do?

____

5. Besides the kidneys, which three other accessory organs form part of the excretory system?

____

USE YOUR MIND

Each year over 6,000 people in the US donate their kidneys, and in the last 10 years, about 1,200 people have donated their kidneys to a complete strangers. Approximately 15% of all anonymous donors were Orthodox Jews, even though they only make up 0.2% of the US population. What are some of the reasons you think they would do this?

____

____

SECTION 2.3: EXPAND YOUR MIND

SEEING THE PLAN AND THE PURPOSE

When learning about the anatomy and physiology of the body, it's easy to get lost in the details. If we do, we may fail to appreciate the brilliant plan and purpose of each of the organs Hashem gave us in order to carry out the trillions of actions needed to survive every moment.

Take a minute to focus on the act of breathing and the wisdom of its Designer that such an act reveals. Imagine there was a sick person who required very specific conditions in the air he breathed. It was essential that the man had warm, moist air that was clean of foreign materials, or else he would be in great danger of becoming seriously ill. The man was very wealthy, so he paid the best doctors and engineers to design an air-filtration system in his home.

As the air entered from the outside, it went through a series of heating chambers, where the cool air from the outside was heated to the optimal temperature before entering the house. As it traveled through those heating chambers, there were a number of filters that the air had to pass through that captured the dust, debris, and bacteria from the air. Then, every hour or so, the entire system was cleaned automatically to prevent a backup of waste.

In addition to the filtration system, the design team hired for this job was really smart and creative. As a surprise, they made a special set of strings that vibrated as air passed through them. The wealthy man was then given a remote control so he could request virtually any type of music he wanted. The air speed would be regulated to speed up or slow down as it passed through the system of strings to make pleasant sounds and beautiful music.

By now you have probably realized that you yourself have a much more sophisticated system—that you received for free! It is made up of your nasal cavities, voice box, and trachea—and it operates 24 hours a day, never raises your electric bill, and runs for 120 years without needing repairs. How much do you think that wealthy man would have paid for his custom-made air-heating and filtration system that played whichever tunes he wanted to hear? I am sure it would have cost a small fortune. If that ill man had been poor, however, and a wealthy friend had showed him great compassion by paying for that lifesaving system, the ill man would surely have been forever grateful, and showered praises on his kind and generous friend. If so, how much more thanks and praise do we owe to our Creator for giving us our own personal air-heating and filtration system that also enables us to speak words and sing songs? In appreciation for this wondrous gift from our Creator, it is the least we can do to use our words to give Him *nachas*, by saying nice things to other people, speaking words of Torah, and singing songs of praise to Hashem!

SECTION 2.3: EXPLORE CHAZAL

TORAS EMES

The Torah teaches us to "be extremely protective of [our] lives" (Devarim 4:15). What exactly is this verse telling us, and in what ways are we supposed to be protecting our lives?

Rabbi Shimshon Raphael Hirsch explains this mitzvah to mean that you may not in any way weaken your health or shorten your life. He writes Only if the body is healthy is it an efficient vehicle for the soul's activity… That's why you should avoid anything which might harm your health… And the law asks you to be even more careful in avoiding danger to life and limb than in the avoidance of other transgressions (Horeb, chapter 62, section 428). Similarly, the Rambam explains that maintaining a healthy and sound body is part of following Hashem's ways, since one cannot understand or have any knowledge of the Creator if one is ill. Therefore, one must avoid that which harms the body and accustom oneself to that which helps the body become stronger (Mishneh Torah, Hilchos De'os 4:1).

Under Jewish law, numerous activities are forbidden due to the health risks involved. These include: walking under a shaky stone wall, avoiding the consumption of human sweat (mentioned earlier), not keeping a wild dog as a house pet, or allowing other unsafe conditions around the home. Regarding the Torah obligation to protect one's life, there has been extensive discussion regarding the permissibility of smoking cigarettes. There is a long list of gedolei Yisrael who forbid smoking, due to the great harm it causes the body. This list includes: Rabbi Shmuel Auerbach, zt"l; Rabbi Yosef Shalom Elyashiv, zt"l; Rabbi Aharon Leib Shteinman, zt"l; Rabbi Shmuel HaLevi Wosner, zt"l; Rabbi Ovadia Yosef, zt"l; and Rabbi Chaim Kanievsky, shlit"a; Rabbi Moshe Sternbuch, shlit"a; and Rabbi Asher Weiss, shlit"a, to name a few.

One of the most famous rabbinic teshuvos on the issue of smoking was from Rabbi Moshe Feinstein, zt"l, who ruled leniently on this issue following the 1964 Surgeon General's report on the health issues involved with smoking (Igros Moshe, Yoreh De'ah, vol. 2:49). While Rav Moshe said it was clearly forbidden to smart smoking, he did not require one who already smoked to quit even though it was certainly advisable to do so. Many rabbis today are of the opinion that Rabbi Feinstein's ruling was based on the 1964 Surgeon General's report, and that if Rabbi Feinstein had been aware of the recent finding that

show smoking is much more dangerous to a person's health than they had originally thought, he would not have ruled leniently in the matter. Among those who share this view is the Rabbinical Council of America, who ruled in 2006 that they believe Rabbi Feinstein would have agreed with their decision that smoking is "clearly and unquestionably forbidden by halachah and that this should be made known to all who care about the Torah and their health." Similarly, Rabbi Dovid Feinstein, shlit"a, was quoted as saying that with the current knowledge of the harm smoking causes, it is obvious that had his father, Rav Moshe, still been alive today, he would have prohibited smoking outright, as his father's dispensation was based only on the former "fact" that smoking endangered only a small percentage of smokers!

Despite the severity of the prohibition to smoke, the Rabbinical Council of America concluded its ruling with the following: "People who smoke are not, G-d forbid, doing so in an attempt to flout halachah. In fact, most would dearly wish to quit, but shedding an addiction is no simple matter. While it is important to make clear that halachah prohibits smoking, it is also important not to condemn those who struggle with this issue. Rather we must offer our full help and support to aid them in their quest for physical and spiritual health."

EXPERIMENT:

DIAPHRAGM SIMULATION

by Science Sparks

Corresponds to Section 2.1

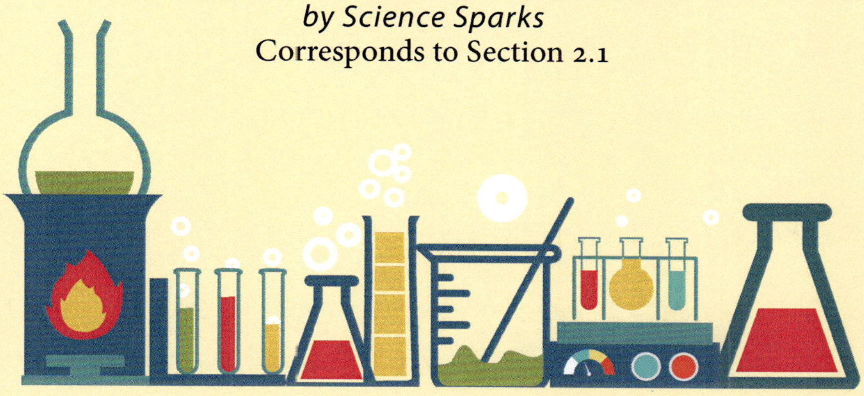

Introduction

When you inhale, your diaphragm, a band of muscles at the bottom of your chest cavity, contracts. When your diaphragm contracts, it flattens and pulls downward. As it pulls downward, the volume of your chest cavity increases, and the pressure inside it decreases. The pressure of the air outside your body is then greater than the pressure inside the chest cavity, so air rushes in through your mouth and nose. When you exhale, your diaphragm relaxes and returns to its original shape, decreasing the volume and increasing the air pressure inside your chest cavity. This forces air out of your lungs.

The model below simulates the contraction of the diaphragm (portion of balloon attached to the bottom of the cut plastic bottle) and the resulting rush of air into the lungs (the balloon inside the bottle). The increase in pressure inside the bottle causes air to rush into the balloon, just as the increase in pressure in your chest cavity causes air to rush inside your body when your diaphragm contracts.

Materials:

- A plastic bottle
- A straw
- An elastic band
- Scissors
- 2 balloons
- Play dough

Procedure:

1. Carefully cut your bottle to about half the size.

2. Tie a knot in one end of one balloon and cut off the larger end.

3. Stretch the balloon around the bottom of your plastic bottle.

4. Put a straw in the neck of the other balloon and secure tightly with the rubber band but not so much that you crush the straw. The air must flow through, so test it with a little blow through the straw to see if the balloon inflates.

5. Put the straw and the balloon into the neck of the bottle and secure with the play dough to make a seal around the bottle – make sure that again, you don't crush the straw.

6. Hold the bottle and pull the knot of the balloon at the bottom. What happens?

You should find that the balloon inside the bottle inflates, and as you let go the balloon deflates.

Why Does This Happen?

As the knotted balloon is pulled it creates more space inside the bottle. Air then comes down the straw and fills the balloon with some air to fill the space! When you let go of the knot the space no longer exists, so the air from the balloon is expelled making it deflate.

Inside the lungs are a network of tubes which allow air to pass though. Air is warmed, moistened and filtered as it travels through the mouth and nasal passages. It then passes through a network of tubes, eventually reaching tiny sacs called alveoli which are where gas exchange occurs.

How Do Lungs Work?

This fake lung demonstrates how our lungs work. Air is taken in through the mouth and nose, passes down the windpipe and into our lungs. The diaphragm at the bottom of our chest moves down to create more space. As we breathe out the diaphragm raises again. The knotted balloon represents the diaphragm and the balloon inside the container the lung.

EXPERIMENT:

LUNG CAPACITY

Corresponds to Section 2.2

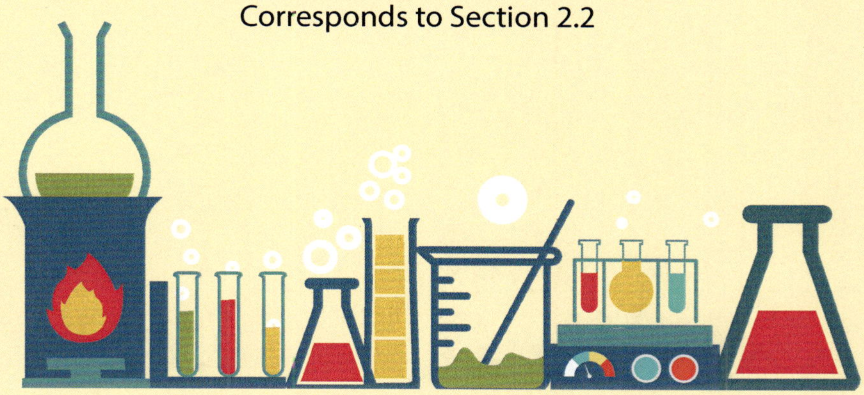

Introduction

The amount of air that you move in and out of your lungs while breathing normally is called **tidal volume**. "Volume" refers to the amount of space that an object or substance occupies, and in this case refers to the volume of air in the lungs. The tidal volume of air provides enough oxygen for a person who is resting, but it is possible to inhale and exhale more forcefully and bring more air into the lungs than during a normal breath. The maximum amount of air that can move in and out of the lungs is called the **vital capacity**.

In this activity, you will be measuring the vital capacity of members of the class. If you have any breathing difficulties (asthma or another condition), you should not participate. You will be using the diameter of a balloon and the graph below to estimate your capacities.

Materials

- 4 round balloons
- a metric ruler

Procedure

1. Working in groups of 4, record data on the height, age, and typical activity level of each group member. For students who exercise or play sports, record their activity level as "high"; for those who sometimes do those things, write "medium"; if someone almost never engages in physical activity, mark down "low."

2. Measure vital capacity: Pull a balloon several times to stretch it out. Take a deep breath and then exhale into the balloon. Pinch the end of the balloon and measure its diameter in centimeters.

3. Record this number on the data table below.

4. Convert the diameter to lung volume using the graph below, and record this number on your table.

5. Estimate each group member's vital capacity by measuring balloon diameter and converting it to lung volume using the graph below.

Data Table

	Subject Data			Vital Capacity	
	Height	Age	Activity Level low, medium,) (high	Balloon Diameter	Lung Volume (from graph)
Subject 1					
Subject 2					
Subject 3					
Subject 4					

Measure and Graph

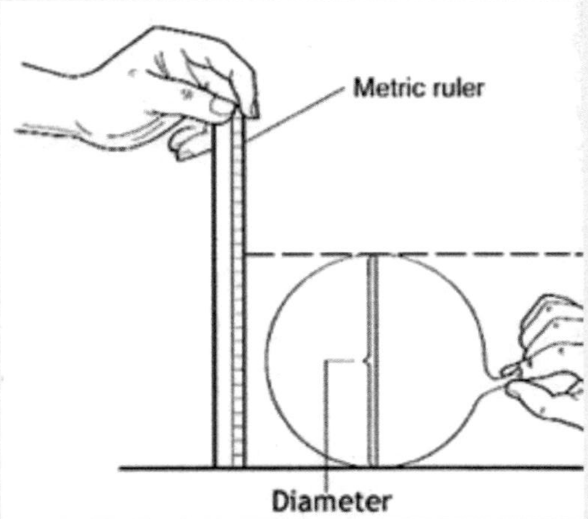

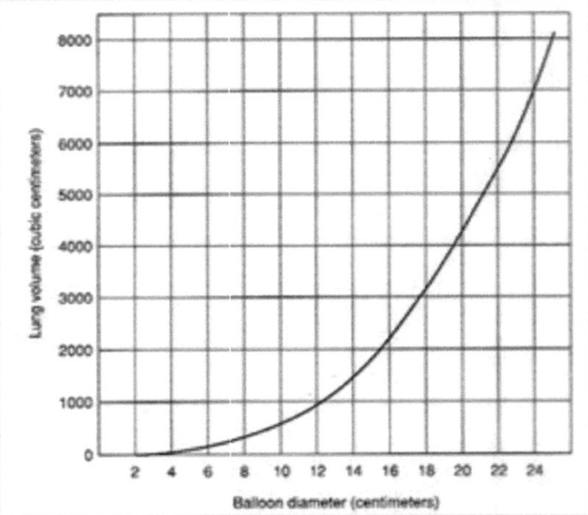

Questions

1. Compare your subjects' vital capacity and variables listed (height, age, and activity level). Based on your limited data, which factor(s) had the greatest impact on vital capacity?

____

2. How might an athlete's vital capacity compare to that of a non-athlete? Explain your reasoning.

____

3. Was your vital capacity higher or lower than the average* for your age and height?

____

* To determine the average vital capacity for your age, use the following formulas:

Male: vital capacity (ml) = (27.63 - 0.112 × age) × height (cm)

Female: vital capacity (ml) = (21.78 - 0.101 × age) × height (cm)

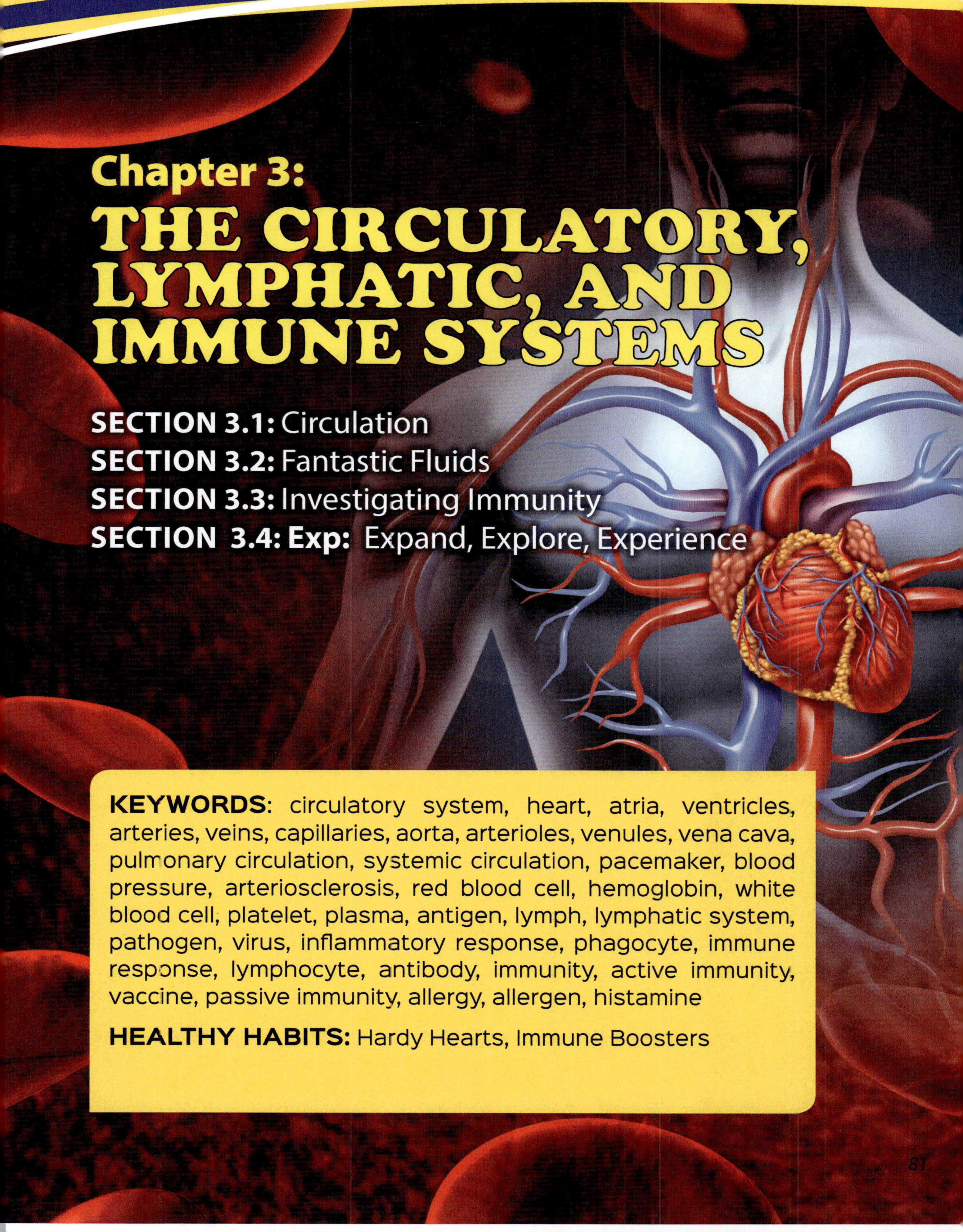

Chapter 3:

THE CIRCULATORY, LYMPHATIC, AND IMMUNE SYSTEMS

SECTION 3.1: Circulation
SECTION 3.2: Fantastic Fluids
SECTION 3.3: Investigating Immunity
SECTION 3.4: Exp: Expand, Explore, Experience

KEYWORDS: circulatory system, heart, atria, ventricles, arteries, veins, capillaries, aorta, arterioles, venules, vena cava, pulmonary circulation, systemic circulation, pacemaker, blood pressure, arteriosclerosis, red blood cell, hemoglobin, white blood cell, platelet, plasma, antigen, lymph, lymphatic system, pathogen, virus, inflammatory response, phagocyte, immune response, lymphocyte, antibody, immunity, active immunity, vaccine, passive immunity, allergy, allergen, histamine

HEALTHY HABITS: Hardy Hearts, Immune Boosters

SECTION 3.1: CIRCULATION

You may take it for granted because it is so common, but it's a real blessing to have stores nearby where you can buy virtually anything you need. Whether it's food, furniture, or building supplies, you have access to the resources of our world when you need them.

Imagine what takes place in order to keep all those stores stocked with goods. Each product comes from a different place. There are farms and factories all over the country, and the world, which produce the items we all need. After being harvested or manufactured and then packaged, the goods are sent in delivery trucks. Those trucks travel along highways, smaller roads, and finally local city streets before reaching their destinations. Designing highways and roads is no simple task; there are many hardworking engineers and city workers who deserve our thanks for such organized and efficient transportation routes.

The truth is that within your own body, you have your own network of highways and roads that was designed to provide a means of transporting vital goods between all of the different parts of the "cellular city" known as your body. We call this transportation network the cardiovascular or **circulatory system**, and it is made up of the heart, blood vessels, and blood. You will be learning about that in this chapter. In addition to directing the flow of traffic that carries nutrients via the blood to the cells of your body, the circulatory system helps fight illness by delivering disease-fighting cells, which are capable of killing microorganisms that could make you sick.

This is one of the first "driverless trucks," and a more accurate analogy to the transportation "vehicles" Hashem put inside your body.

While the analogy of delivery trucks carrying nutrients on roadways to the cellular cities throughout your body is a good one, the truth is that the circulatory system does much more than that. It also helps perform the function of waste management by carrying the garbage disposed of by the cells, such as carbon dioxide produced as a by-product of the breakdown of sugars, to the kidneys, where the wastes are filtered out and removed. It is a remarkable fact that the same roads that carry nutrients to the cells, also carry wastes away from them for removal. The different components of the blood stay separated from each other long enough for them to be delivered to their designated addresses.

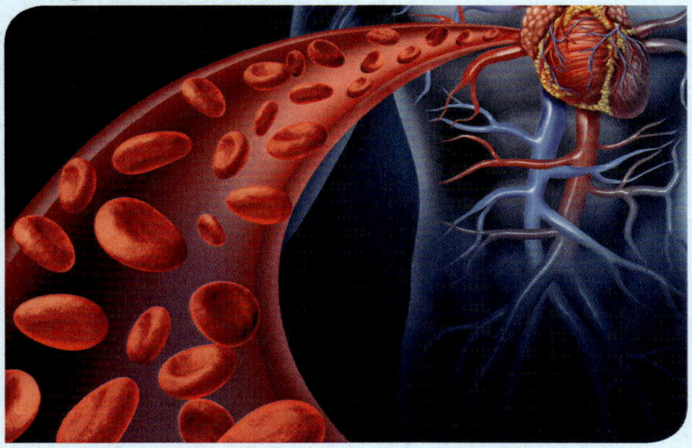

How long do you think the roadways Hashem put inside your body are? While most of the blood vessels are very small in width (about 8 microns or 0.008 millimeters), if you were to lay them all down end to end to measure their length, they would stretch 100,000 miles (161,000 kilometers)! How long is that? It's four times around the world and back! While modern roadway engineering is impressive, it cannot compare to the Divine Design that fits 100,000 miles of transportation lanes inside of every person's body.

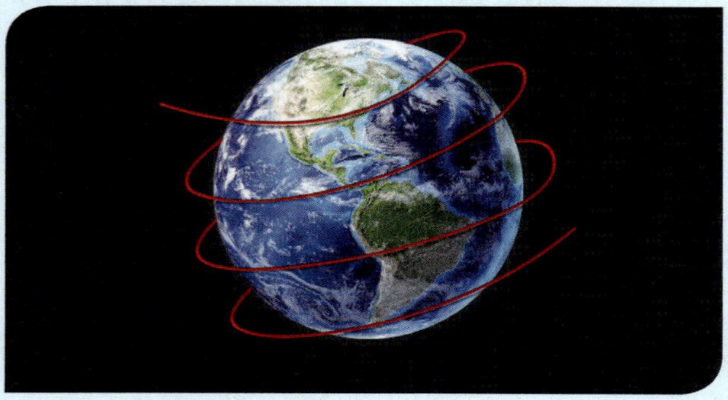

Structure of the Heart

The **heart** is the largest and most prominent organ in the circulatory system. It is a hollow, muscular pump that pushes blood throughout the body. Every time the heart beats, it pumps blood around the body by way of blood vessels. In light of its vital role, the heart is well protected, located in the center of your chest behind the sternum, a flat bone that is 6 inches (15 centimeters) long and 1 inch (2.5 centimeters) wide, and surrounded by the rib cage.

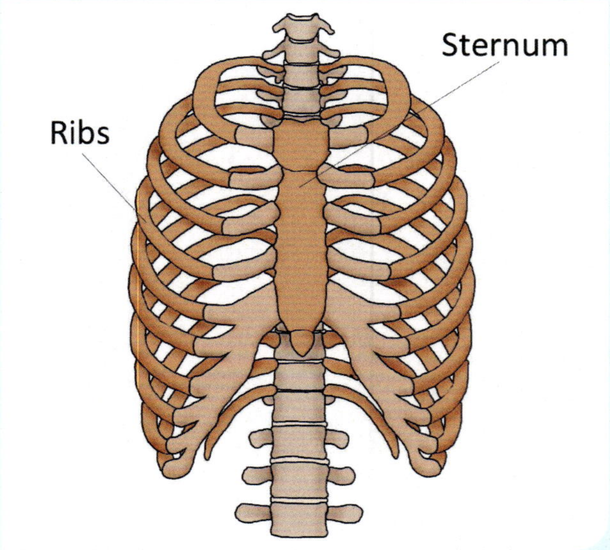

The heart is a remarkably powerful muscle that contracts continuously without ever getting tired. It pumps about 100,000 times a day, or 35 million times a year, and sends about 2,000 gallons (7,500 liters) of blood around your body each day. It is not that you have this much blood in your body. The average adult has about 1.5 gallons (5.5 liters) of blood in their body at any given time, but that blood is pumped around and around all the time so there is a constant supply of oxygen and nutrients for your cells. The heart is such an efficient pump that over the course of the average person's life span, the amount of blood pumped by the heart could fill over 80 Olympic-size swimming pools!

This is about the quantity of blood that flows through your body every year!

There are two sides to the heart, the right and the left. The two sides are completely separated from each other by way of a wall of tissue called the septum. Each side is also divided into two parts, an upper and a lower one, which are separate compartments or chambers. As such, the heart has a total of four chambers. The two upper chambers are called the **atria** (ey-tree-uhm; singular: atrium), and they receive all of the blood coming into the heart. The lower chambers are called **ventricles** (ven-tri-kuhls) and they pump blood that enters them from the atria out of the heart to the rest of the body.

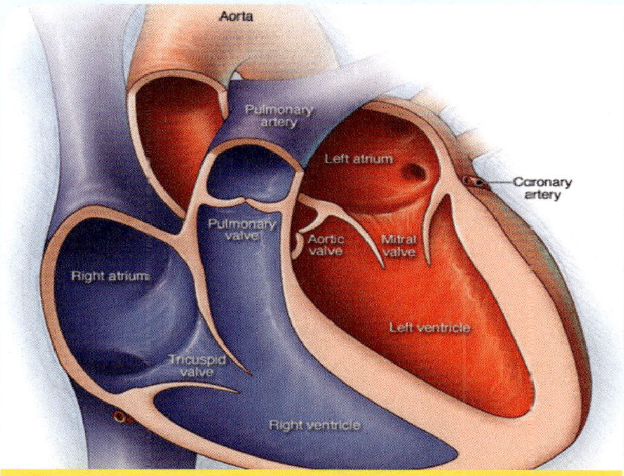

Keep in mind that when looking at a diagram of the heart, the right side of the heart is on the left and the left side is on the right.

It is important for the health of the body that blood flows in only one direction. A backflow of blood would lower the pressure inside of arteries and veins, and having a separate second set of blood vessels running in the opposite direction would require a much larger body, a second heart, and be much less efficient in general. In order to arrange a successful, efficient, one-way, one-lane system, Hashem designed the heart with valves—flaps of tissue that prevent blood from flowing backwards. The valves are located between the atria and ventricles, and between the ventricles and the large blood vessels that carry blood away from the heart.

Pumping Action of the Heart

If you put your hand on your heart or just pay attention very closely (this is easier after exercise), you will feel this powerful pump in action, beating over and over to push blood around your body. Let's follow a single heartbeat to find out how this amazing organ works.

The first thing you need to know is that there are two phases of action. In the first phase, the heart muscles relax and blood flows into all its chambers. In the second phase, the heart muscles contract and push or pump blood forward. These two phases take place in the upper chambers (the atria) and the lower chambers (the ventricles) of the heart. The atria and ventricles take turns relaxing and contracting, which, as we will see, enables a one-way flow of blood.

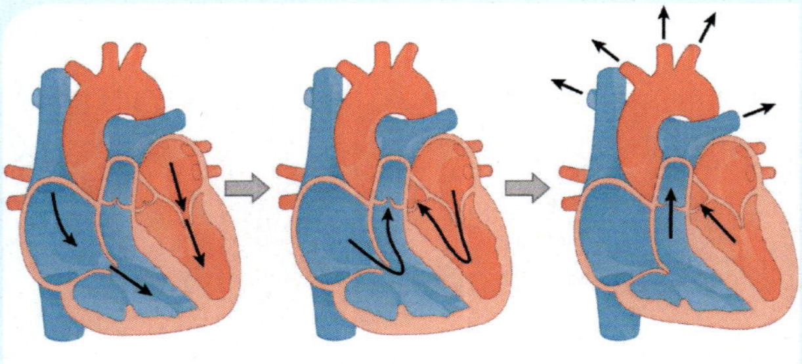

The relaxed phase of the atria or ventricles is called diastole (dahy-as-tl-ee) and the contractive phase is called systole (sis-tuh-lee).

When the heart muscles of the atria contract, the ventricles relax and blood is pushed out of the atria into the ventricles, filling them up. Once full, the ventricles contract, pushing the blood out into the arteries. When the ventricles contract, the atria relax, so that as blood is being pushed out of the ventricles, the atria are being filled up and getting ready for the next beat.

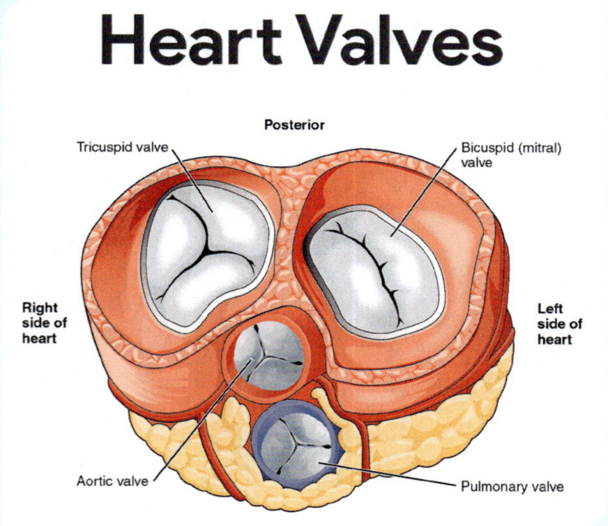

The pumping action of the heart can be heard using a stethoscope. It sounds something like "*lub-dup*." This *lub-dup* sound is not the sound of muscles contracting; it's the sound of the valves that separate the atria and ventricles, and the valves that separate the ventricles and blood vessels shutting forcefully. When the ventricles contract and send blood out into the blood vessels of the body, the valves separating the atria and ventricles shut to prevent a backflow—that is the *lub* sound that you hear. Once the ventricles have pushed their contents into the blood vessels, they relax and cause the valves that separate between them and the blood vessels to shut, causing the *dup* sound. With your newfound knowledge, you can better appreciate each and every *lub-dup* taking place within you. You'll also probably really impress your parents if you ever try explaining this to them, since most people don't really understand the process of how blood flows through their heart.

Types of Blood Vessels

Before we continue exploring how the heart works, we need to discuss the types of tubes that run throughout your body and carry blood to and from your heart. There are two main types of blood vessels: **arteries** (ahr-tuh-rees), which carry blood away from the heart, and **veins** (veyns), which carry blood back toward the heart. This system was designed so that there is a one-way flow of traffic inside the body ensuring everything gets where it needs to go without getting mixed together.

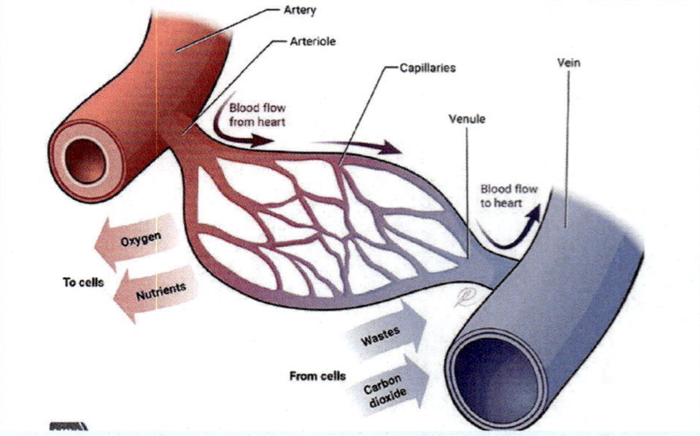

layer is comprised of muscles that contract and relax when you need more oxygen and blood (such as during exercise), and the outer layer is formed from sturdy yet flexible tissue, called connective tissue. This layer gives the artery support and protection.

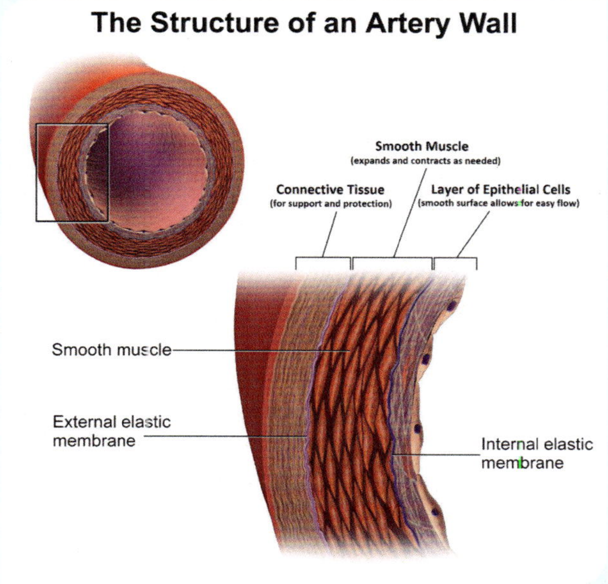

Arteries are much thicker than veins since they must withstand much greater blood pressure than veins. You'll hear more about that soon. In both arteries and veins, however, their three-layer structure makes your blood vessels. A third type of blood vessels is called **capillaries** (kap-uh-ler-ees). Those are the smallest type of blood vessels, just one cell thick; they are the sites where substances are exchanged between the blood cells and the body's cells.

There are two main arteries that carry blood away from the heart. One, called the pulmonary artery, leaves the right ventricle and carries blood to the lungs. The other, called the **aorta** (et-awr-tuh), is the largest artery in the body. It carries blood pumped by the left ventricle out toward the rest of the body. The aorta is the "superhighway" of arteries since it has many "off-ramps" consisting of smaller arteries leading to the various organs of the body. The connecting arteries get progressively narrower and narrower until the narrowest ones, the **arterioles** (ar-te-ri-oles), connect to capillaries, which supply cells with oxygen, pick up carbon dioxide waste, and exchange other substances.

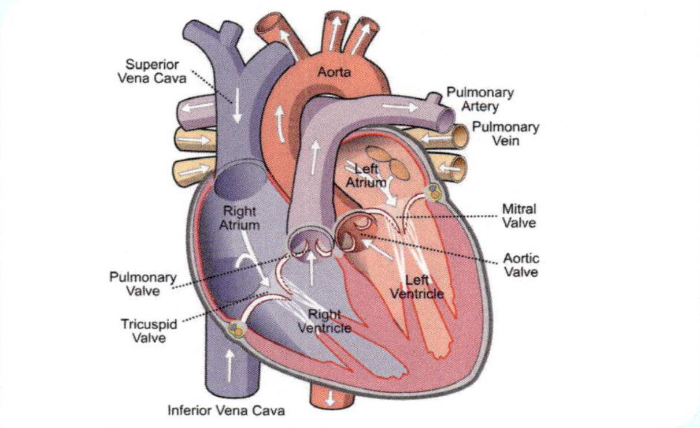

Those same capillaries are also connected to the smallest veins, called **venules** (ven-yools). It is in those venules that a different journey begins—the journey of bringing both deoxygenated blood (blood low in oxygen, since it has already delivered most of its oxygen content to the cells of the body) and carbon dioxide back to the heart. The venules lead into progressively wider and wider veins that eventually bring deoxygenated blood back to the heart through two large veins called the **vena cava** (vee-nuh key-vuh). The first vena cava is called the superior vena cava because it is located at top of the heart, and veins from the upper body lead into it. The second vena cava is called the inferior vena cava because it is located below the heart, and al the veins from the lower part of the body lead into it.

Two-Circuit System

As you learned above, the right and left ventricles both pump blood away from the heart. What you may not have realized, however, is that each side pumps blood to different places. This creates two separate systems of circulation, referred to as pulmonary circulation and systemic circulation, or the pulmonary and systemic circuits.

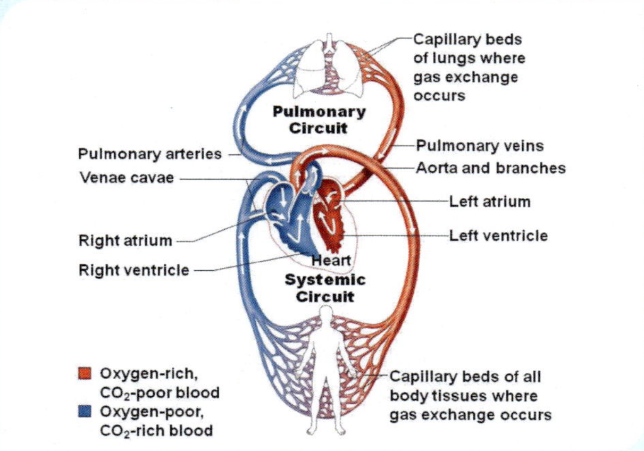

Pulmonary (puhl-muh-ner-ee) **circulation** carries blood from the heart to the lungs, where it picks up oxygen and dumps off carbon dioxide, and then brings the blood back to the heart. **Systemic** (si-stem-ik) **circulation** takes the oxygenated blood brought from the lungs, pumps it to the rest of the body, where it supplies the cells of the body with oxygen, and takes away carbon dioxide waste, and then brings it back to the heart to start the cycle again. To better understand blood flow, let's follow the journey of blood through the body.

Let's take a long, one-minute trip… At any given time, it takes a drop of blood only about 1 minute to make a full trip around both the pulmonary and systemic circuits and come back to where it started. We will be starting out in your foot, let's say with the blood in your big toe.

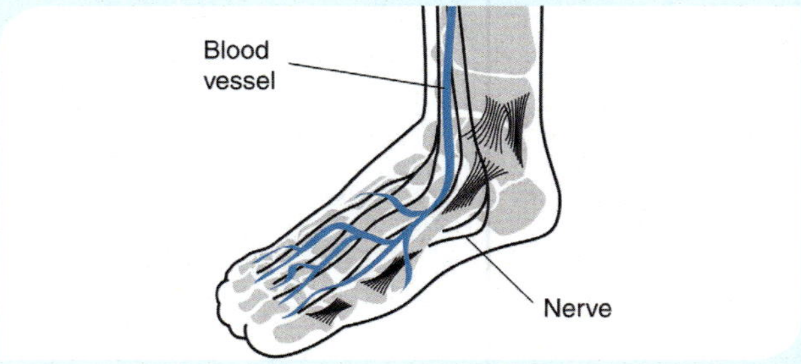

The arteries that brought the blood there have supplied the cells of your big toe with oxygen, and picked up the carbon dioxide wastes they gave off, for removal. Now the big toe veins send the deoxygenated blood back up toward the heart.

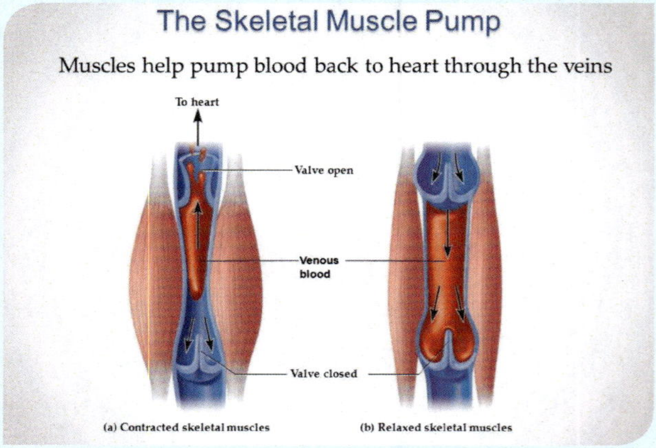

In general, the force or pressure of blood flow is much lower in narrow veins than in wide arteries. This is especially true in the toes, since they're so far away from the heart that the force of the heart's pumping action is much less. To help get the blood from the toes to the heart, Hashem put many veins right next to the muscles attached to your bones. That way, when you walk and your leg muscles automatically contract, the veins get a much-needed squeeze that helps push the blood back up to the heart.

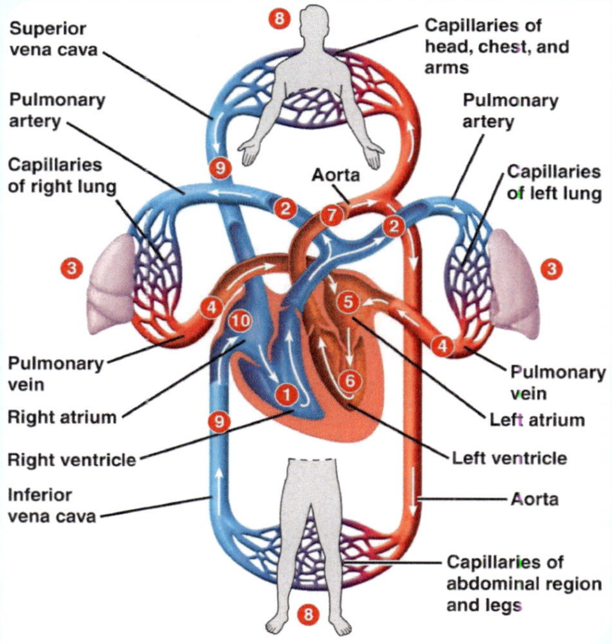

Veins from the toes lead to wider veins in the legs, which eventually lead into the inferior vena cava, from where blood enters the heart. At this point, the blood has little oxygen left and contains a lot of carbon dioxide. It flows into and fills up the right atrium, which then contracts and pushes the blood into the right ventricle. From here, the right ventricle pumps the blood into an artery leading to the lungs, called the pulmonary artery. The pulmonary artery carries the blood to the lungs to pick up oxygen.

The large arteries in the lungs become narrower and narrowed until they turn into capillaries that are so narrow and thin that oxygen diffuses right into them and carbon dioxide diffuses out. You may recall that diffusion refers to the flow a substance from a higher concentration to a lower concentration. Because air coming into the lungs when we breathe has a higher concentration of oxygen than in the blood, oxygen diffuses into the blood. Similarly, because the blood has a higher concentration of carbon dioxide than the air coming into the lungs, carbon dioxide diffuses out into the lungs and is expelled from the body when you exhale.

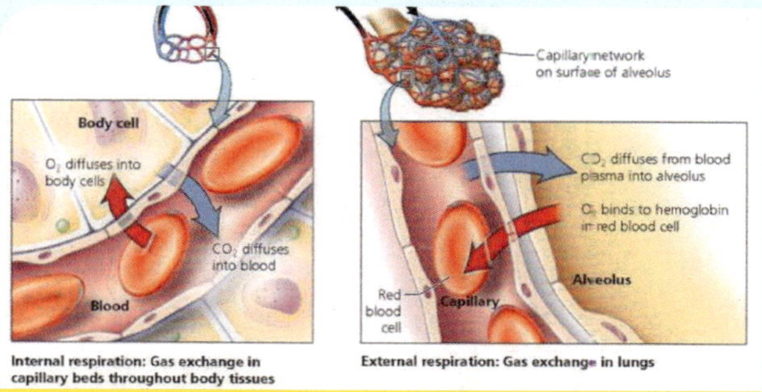

Diffusion of oxygen and carbon dioxide in the body

After the blood fills up with oxygen, it continues to flow inside pulmonary veins, which bring blood back to the heart, filling up the left atrium. The left atrium then contracts, pushing the blood into the left ventricle. From here, the left ventricle contracts and pumps the oxygen-rich blood out of the aorta. The aorta branches out into narrower arteries, one of which brings oxygen-rich blood back to your big toe, thus completing the journey.

The heart is a great gift and a sign that Hashem loves us a lot!

Take a moment to think about how this cycle is taking place inside you, every minute of your life without any conscious thought. It is an awesome feat of engineering that seems almost unbelievable. I personally think that this incredible invention deserves a big round of applause from all the inhabitants of the whole world who have one of these marvelous machines—but at the least we should say thank You to Hashem once in a while, for the remarkable heart and blood-transportation system He gave us.

Pulse

The easiest way to see, or to be precise, feel your circulatory system in action is to lightly touch the inside of your wrist. What you are feeling is the rising and falling of the artery that supplies blood to your hand. That is called your pulse, and it is caused by the expanding and contracting of the artery wall. When the heart's left ventricle contracts, it uses great force to send a gush of blood out of the aorta and through the arteries around your entire body (except for your lungs, which are supplied by the right ventricle). As this blood travels through the arteries, it pushes on the artery walls and makes them expand. After the stream of blood passes, the artery walls relax and shrink.

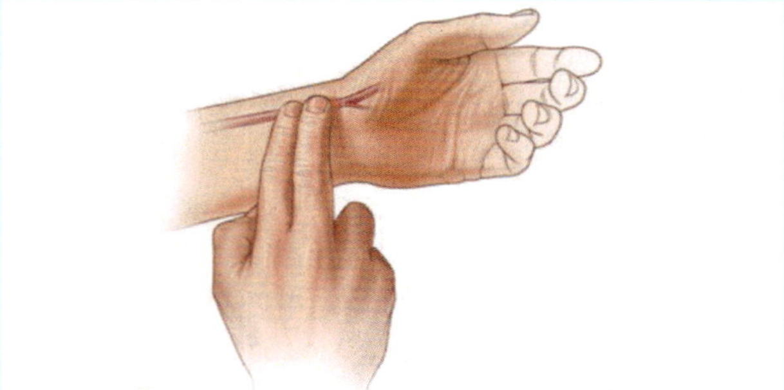

beats, since each beat of the heart causes one gush of blood to be sent out of the aorta and to the artery supplying your wrist with blood—causing it to pulse. Knowing the pulse rate is useful since it is one of the ways to determine the health of the heart. A pulse of 60 to 100 beats per minute is normal for an adult. In general, a lower pulse means that the heart is healthier and a higher one is a sign that the heart is working harder than normal to supply the body with blood. That is often a sign of poor physical fitness.

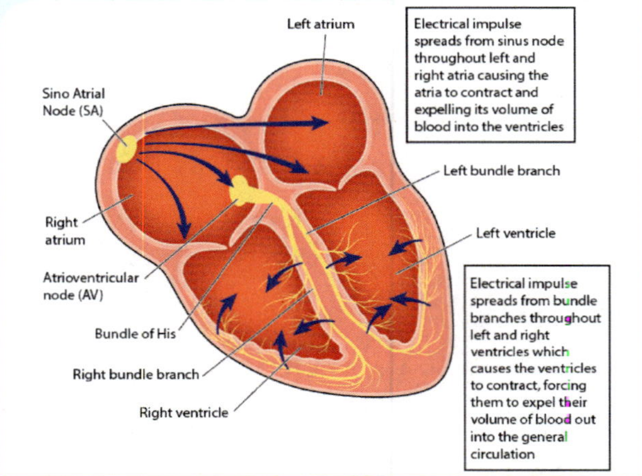

Diagram showing the generation and flow of electricity in the heart by a pacemaker that regulates its contractions

Another fascinating aspect of the heart is how it keeps up its pace of regular beats. There is a special group of heart cells located in the right atrium called the sinoatrial node, commonly referred to as the natural **pacemaker**—not to be confused with artificial pacemakers that are surgically inserted in the body to replace the natural one when it ceases to function normally. The pacemaker sends out electrical signals that trigger the heart muscle to contract. The pacemaker ensures that the heart does not work harder than necessary or less than is necessary, by regulating the pace to speed up the heart when you need more energy and slow it down when you are resting.

Blood Pressure

Another way of measuring the health of the heart and circulatory system is by taking your blood pressure. You may have had your blood pressure taken by a doctor or nurse who wrapped an instrument around your upper arm, pumped air out of it so it squeezed your arm, and then released the air, all while recording the numbers they measured. What they were measuring is what we call your **blood pressure**, which is the pressure or force that blood exerts on the arteries and veins it flows through. Blood pressure is highest in the vessels closest to the heart and decreases with distance. The pressure in vessels furthest from the heart is therefore even less.

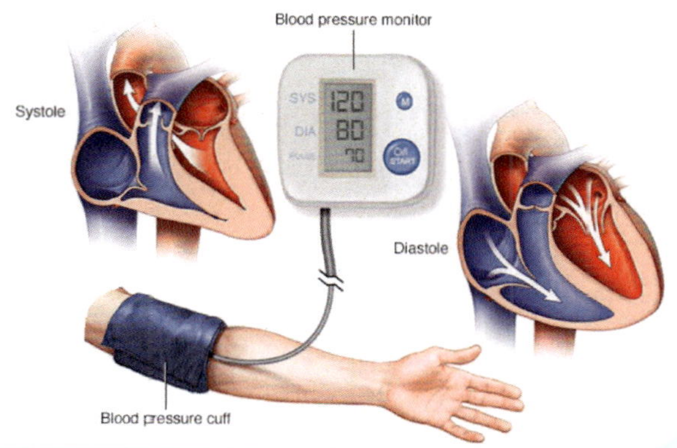

Knowing a person's blood pressure is important because if it is too high (a health situation called hypertension), that can be a sign of heart disease and circulatory system disorders. One common form of heart disease that causes high blood pressure is called **arteriosclerosis** (ahr-teer-ee-oh-skluh-roh-sis). Arteriosclerosis happens when the arteries get clogged up and blocked by a fatty substance called plaque. The blockage in the arteries causes blood pressure to rise, since the blood has less room to move the same amount of blood through it. If untreated, arteriosclerosis can lead to a heart attack, which is why blood pressure is often measured as part of a general physical examination.

Blood Pressure Categories

American Heart Association | American Stroke Association

BLOOD PRESSURE CATEGORY	SYSTOLIC mm Hg (upper number)		DIASTOLIC mm Hg (lower number)
NORMAL	LESS THAN 120	and	LESS THAN 80
ELEVATED	120 – 129	and	LESS THAN 80
HIGH BLOOD PRESSURE (HYPERTENSION) STAGE 1	130 – 139	or	80 – 89
HIGH BLOOD PRESSURE (HYPERTENSION) STAGE 2	140 OR HIGHER	or	90 OR HIGHER
HYPERTENSIVE CRISIS (consult your doctor immediately)	HIGHER THAN 180	and/or	HIGHER THAN 120

There are actually two types of blood pressure that are recorded. The first is called systolic pressure, which is a measure of the blood pressure when the heart's ventricles contract and pump blood into the arteries. It is always higher than the second type of blood pressure, diastolic pressure, which is the pressure measured when the ventricles are relaxed.

HEALTHY HABITS:
Hardy Hearts

Most people are aware of the need to exercise and eat healthy foods in order to keep our hearts working well, but there are other things we can do to make sure our heart stays as hardy as can be. A. Marc Gillinov, MD, in Cleveland Clinic's *Health Essentials*, lists five key things you need to do every day to help your heart work most efficiently.

1. Eat healthy fats, not trans fats (industry-produced fats commonly found in junk food), which are known to clog up arteries and therefore restrict blood flow. Having clogged arteries increases the risk of developing heart disease or having a stroke.

2. Practice good dental hygiene, especially flossing your teeth daily. Studies have shown that bacteria in the mouth can move into the bloodstream and cause inflammation of blood vessels, which can in turn increase the risk of heart disease.

3. Get enough sleep. Researchers believe sleeping too little aggravates existing medical conditions and disrupts normal biological processes, including blood pressure and the inflammatory response.

4. Do not sit for too long at one time. Research has suggested that staying seated for long periods of time is bad for your health, no matter how much exercise you do.

5. Last but certainly not least—avoid secondhand smoke. This is because the chemicals emitted from cigarette smoke promote the development of plaque buildup in the arteries. Studies show that the risk of developing heart disease is about 30% higher for those regularly exposed to secondhand smoke.

SECTION 3.1 REVIEW

1. What are the components of the circulatory system?

____

2. How many chambers are inside the heart, and what are they called?

____

3. List the three types of blood vessels in the body and what they do.

____

4. List the two circuits of blood flow in the body and where each of them carries blood.

____

5. Explain the difference between pulse and blood pressure.

____

USE YOUR MIND

Based on what you have learned in this section on blood flow through the veins, suggest a possible explanation and an easy treatment for someone who spends most of their day sitting down and is suffering from lower leg pains.

____

____

SECTION 3.2: FANTASTIC FLUIDS

In this section you will be learning about the most fantastic fluid in the universe. It carries all of the nutrients that the cells of your body need, as well as a host of other substances that benefit your body in various ways. Blood is much more complicated than it looks and is made up of four different components: red blood cells, white blood cells, platelets, and a fluid we call plasma.

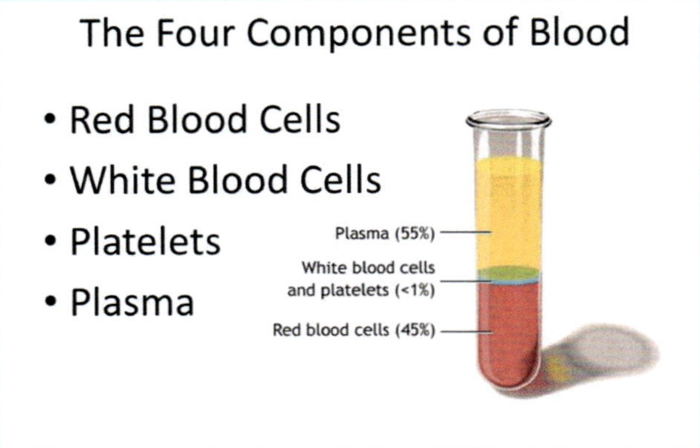

Blood Cells

There are three types of cells found inside the blood: red blood cells, white blood cells, and platelets. Most blood cells are produced inside the marrow—the soft fatty tissues within your bones.

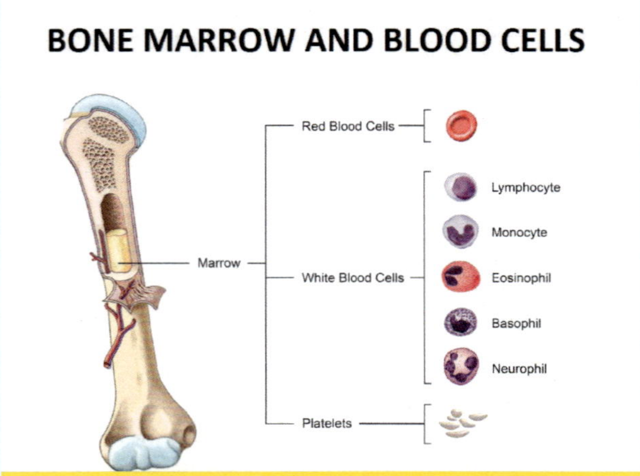

Blood cells form in bone marrow.

Red blood cells look like discs with little indentations in the middle, or bagels without holes. But they're very flexible and can change shape when they need to, which helps them slip into tight places, such as tiny capillaries, where they drop off oxygen to cells that need it. When they mature, red blood cells lose their nuclei, making it impossible to reproduce or repair themselves. They only live about 120 days, and every second, about 2 million red blood cells die. Not to worry, though—your bone marrow miraculously produces new red blood cells at the same rate.

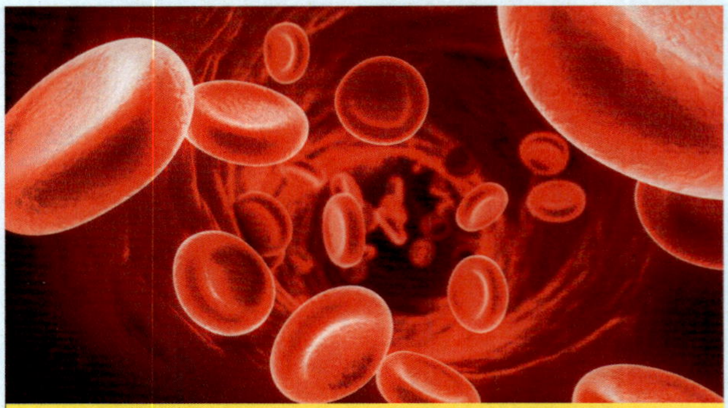

The disc-like shape of red blood cells helps them pass through the smallest blood vessels.

Blood gets its red color because it is mostly made up of red blood cells. These cells get their color from **hemoglobin** (hee-muh-gloh-bin), a protein containing iron that is very good at collecting and transporting oxygen. When oxygen combines with iron, the interaction of them produces a red color. Hemoglobin is what enables your blood to pick up oxygen in the lungs, and then release the oxygen to cells as blood circulates through your body. After releasing its oxygen, hemoglobin then absorbs some of the carbon dioxide from your capillaries and carries it away to be expelled when you exhale. The majority of the carbon dioxide, however, is carried by the plasma content of the blood. When blood is exposed to air, for example when you take a deep breath and blood passes through the lungs, the hemoglobin absorbs lots of oxygen, making it turn a deep red.

White blood cells are your body's soldiers that fight off microscopic invaders.

Along with red blood cells, which make up about 40 to 45% of your blood, there are **white blood cells**, also known as leukocytes, which help fight disease and keep you healthy. Unlike red blood cells, of which there are only one type in each person's body, there are many different types of white blood cells. They are called white blood cells because they have no color and look white under a microscope. Different types of white blood cells destroy parasites, kill bacteria that cause infections, and generally keep your body's immune system working properly. They are much larger than red blood cells and are much less common in the body—there are about 1,000 red blood cells for every 1 white blood cell.

The third type of blood cells are **platelets**, also known as thrombocytes. These are actually small cell fragments that come from giant cells made in bone marrow, which break up into many tiny pieces. Platelets aid in clotting; when a blood vessel breaks or you get a cut, platelets get activated and clump around the site of the wound to help control bleeding. They also release chemicals that trigger the formation of fibrin, a substance that helps create a scab so the wound can start to heal.

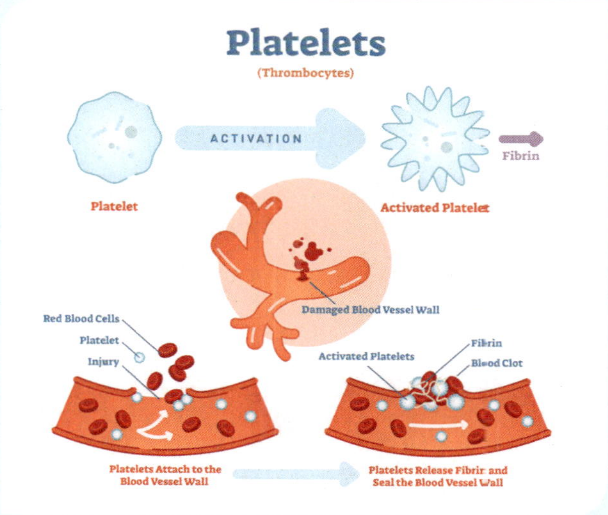

Plasma

All three types of blood cells float in a fluid called plasma. **Plasma** (plaz-muh) is the liquid part of the blood and the place where most nutrients, like sugars, fats, proteins, vitamins, and minerals, are transported. Chemical messengers that help direct activities of the body are also carried in plasma, as well as waste products released by cells, which are taken away in the plasma for elimination from the body.

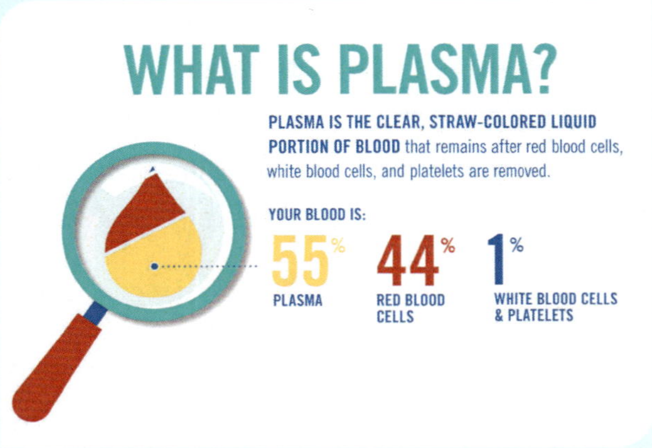

The vast majority of plasma, about 90%, is actually water, and the other 10% is made up of dissolved materials. If viewed on its own, plasma looks yellow, due to the three groups of proteins found inside. Each group of proteins carries out different tasks. One group helps regulate the amount of water in the body, another helps fight disease, and a third group helps form blood clots if the body is wounded.

Blood Types

A blood transfusion refers to the procedure of transferring some of one person's blood to another. This is often necessary after a serious wound or during surgery if a patient loses a lot of blood. Many early attempts at blood transfusions failed. It remained a mystery as to why, until the early 1900s, when Karl Landsteiner, an American physician born in Austria, discovered the trick to successful transfusions.

When Landsteiner began his investigation, he tried mixing blood samples from different people. He noticed that some of the samples mixed smoothly, while in others, the cells clumped together. He immediately recognized it was the clumping of the blood cells that prevented successful transfusions, but he still had to figure out why this was happening. What was different about the blood samples that did not clump?

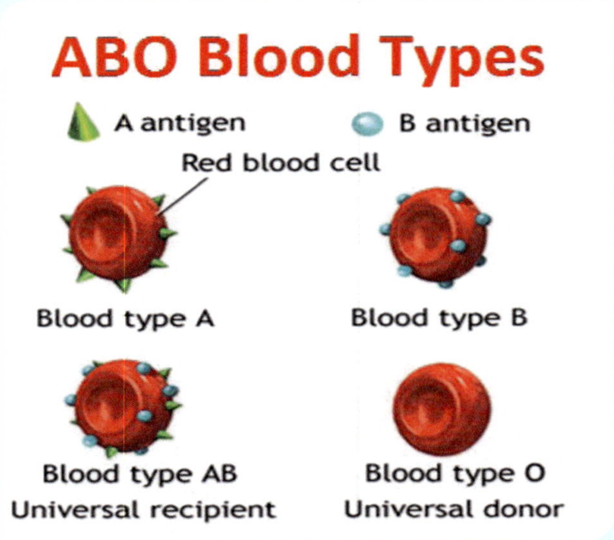

Landsteiner eventually discovered that there are four different types of blood found in humans. He named them types A, B, AB, and O. What makes people's blood different is a type of protein called **antigens**. They are found on the surface of red blood cells and serve as biological markers (essentially "name tags") that tell your body: "This substance is a part of you." People with blood type A have the A antigen, type B have the B antigen, AB have both A and B antigens, and type O have no antigens. Since all living things have antigens, the antigens on the surface of your cells are sort of like their ID card or passport that let the "border control" of your body know they are legal citizens. Due to frequent "illegal immigration" (germs that seek to enter the body and make it ill), all the cells in your body must go through routine security checks to make sure your body systems stay safe.

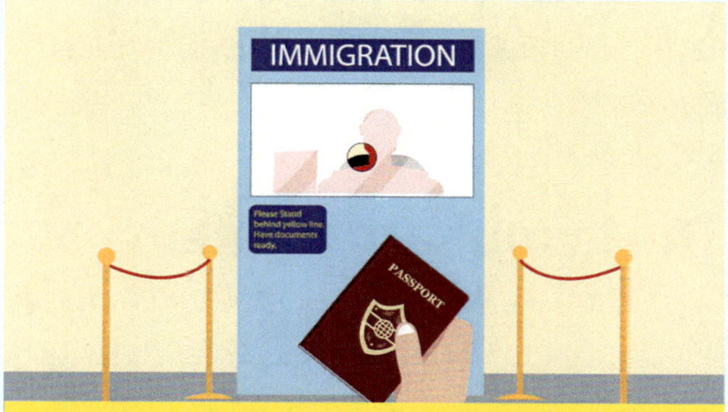

Antigens enable the prevention of dangerous substances from entering the borders of your body, while allowing access to harmless ones.

Your body was designed to detect any foreign cell and actively try to eliminate it. While this normally protects you from harmful invaders, it also makes successful blood transfusions more difficult. When blood with a different antigen is transfused into someone's body, other proteins present in the plasma, called antibodies or clumping proteins anti-A and anti-B (depending if the invader is blood type A or B), latch on to the foreign cells and cause them to clump together in order to target them for destruction. In 1930, Landsteiner received the Nobel Prize for his discovery of blood types because this knowledge enabled the safe and successful blood transfusions that have since saved hundreds of millions of lives.

We now understand that a person with type A blood can only receive a transfusion from someone who also has type A blood, because they both have the same marker molecules on their red blood cells. They can also receive a transfusion from someone who has type O blood, since the latter have no antigen marker molecules on their red blood cells and type A blood will therefore not be rejected. Similarly, someone with type B blood can only receive transfusions from someone with type B or O blood. People with type AB can receive transfusions from all blood types because they have no antibodies out to destroy them in their plasma.

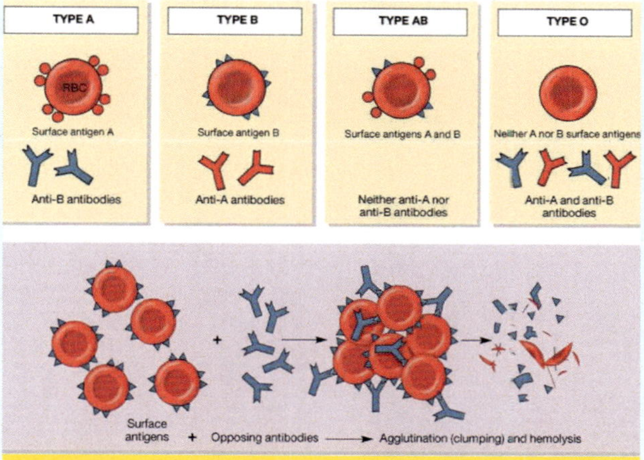

Type A blood has antibodies that attack type B blood antigens. Type B blood has antibodies that attack type A antigens. Type AB has no antibodies, and type O has antibodies that attack both A and B antigens.

For this reason, people with type O blood are called "universal donors." They can help anyone in need, no matter their blood type. But people with type O blood can only receive transfusions from others with type O, since they have antibodies against A and B types in their plasma.

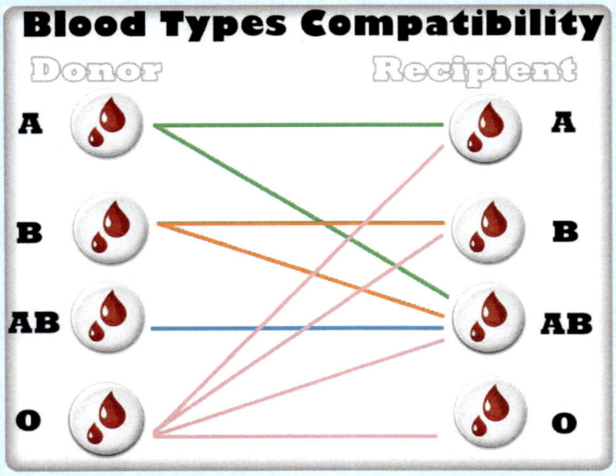

Later on, in 1937, Landsteiner, together with a colleague, Alexander Solomon Wiener, discovered another protein marker on red blood cells, which they called the Rhesus (ree-suhs) or Rh factor. About 85% of people have the Rh factor and are referred to as Rh positive, while 15% of people do not have the Rh marker and are called Rh negative.

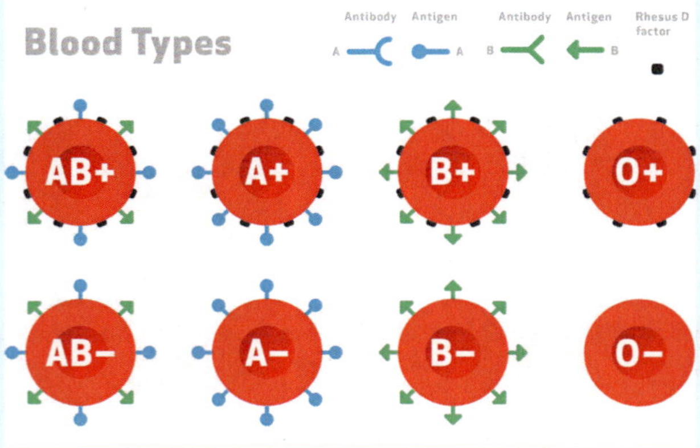

When the presence of the Rh factor is included, there are eight different blood types. Those with the Rh factor are positive (+) and those without the Rh factor are negative (-).

The presence of the Rh marker is important to know because if someone is Rh negative, they do not want to get a blood transfusion from a Rh-positive donor. Even though the blood will not be rejected at the time of the transfusion, the body will eventually develop Rh antibodies that can cause a severe rejection of any Rh-positive blood that is received in the future. For this reason, before any transfusions are done, the blood type and presence of Rh marker is checked to ensure the right match is made and the patient can successfully receive the blood that is being offered.

The Lymphatic System

In addition to blood, there is another fantastic fluid running through your body called **lymph**. It flows through the **lymphatic system**, a series of vein-like vessels that returns fluid to the bloodstream.

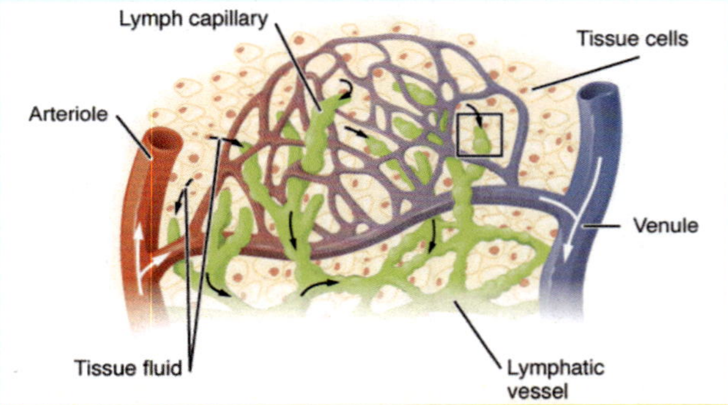

The fluid that bathes cells is called interstitial fluid. Some of it flows into the lymph vessels.

Blood plasma contains a nutrient-containing fluid, called interstitial fluid, which leaks out of small holes in the capillaries as blood flows through them. Interstitial fluid bathes the cells and tissues of your body, providing them with the nutrients and materials they need.

Not all of the lost fluid is absorbed into the cells, however. Some of it is collected in the vessels of the lymphatic system; they act as the body's drainpipes. Once inside the lymph vessels, the clear, watery fluid containing protein molecules, salts, glucose, urea, and other substances is called lymph. Unlike the circulatory system, which is driven by the heart, there is no pump connected to the lymphatic system, so lymph moves much slower than blood.

The upward motion of the interstitial fluid is due to a few factors. Firstly, like some veins, lymph vessels run along muscles whose contraction helps push the fluid along. Secondly, there is a small amount of pushing pressure that comes from the continuous flow of new lymph coming into the system. To prevent backflow, meaning to ensure a one-way flow of fluid, there are also numerous valves located throughout the lymphatic system. Eventually the lymph makes its way up to large veins in the chest, where the fluid reenters the circulatory system and becomes a component of the blood plasma again. In this way, the lymphatic system helps recycle fluids in the body, but also plays another key role in keeping you healthy.

The lymphatic system uses small knobs of tissue called lymph nodes to filter out germs that could make you sick. The nodes can do this because they contain special types of white blood cells called lymphocytes. Lymphocytes fight off infections by producing chemicals that help your body destroy and then remove foreign invaders, such as bacteria, viruses, fungi, parasites, and toxic chemicals.

A person generally can't feel their lymph nodes, but when the body is in the process of fighting off an infection, they can be felt in the neck. That is often one of the first places a doctor will feel or check if an infection is suspected, since the lymph nodes swell from all the filtering and "fighting" going on in there.

SECTION 3.2 REVIEW

1. What is plasma and what is it made up of?

___

2. What is the function of red blood cells?

___

3. What are platelets and what function do they serve?

___

4. Why can some blood types mix and others cannot?

___

5. What is the function of the lymphatic system?

___

USE YOUR MIND

Keeping the Rhesus (Rh) factor in mind, which blood type is the hardest to find a blood donor for?

___

___

SECTION 3.3: INVESIGATING IMMUNITY

It is not only the lymph nodes that help keep you healthy. You actually have an entire arsenal of defense forces available to battle the microorganisms that get into your body and have the potential to make you sick. The four major groups of invaders that can make people sick are collectively referred to as germs or pathogens (path-uh-jens). The most common kinds are bacteria and viruses, but fungi and protists can also cause illness.

Type of pathogen	Description	Human diseases caused by pathogens of that type
Bacteria *Escherichia coli* 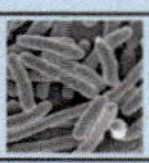	Single-celled organisms without a nucleus	Strep throat, staph infections, tuberculosis, food poisoning, tetanus, pneumonia, syphilis
Viruses *Herpes simplex* 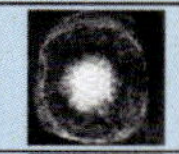	Thread-like particles that reproduce by taking over living cells	Common cold, flu, genital herpes, cold sores, measles, AIDS, genital warts, chiken pox, small pox
Fungi *Death cap mushroom* 	Simple organisms, including mushrooms and yeasts, that grow as single cells or thread like filaments	Ringworm, athlete's foot, tinea, candidiasis, histoplasmosis, mushroom poisoning
Protozoa *Giardia lamblia* 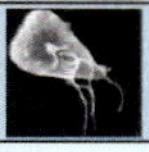	Single-celled organism with a nucleus	Malaria, "traveler's diarrhea" giardiasis, trypanosomiasis ("sleeping sickness")

Bacteria

Bacteria are single-celled organisms and they are the most numerous of all microscopic life. There are an estimated 40 million bacteria living in 1 gram of soil, and about 39 trillion living inside your body! Fortunately, the vast majority of bacteria are harmless and many of the ones living inside your body actually help you, such as those found in the large intestines that make vitamins you need. Some bacteria can cause damage and are capable of causing a wide variety of diseases, including strep throat, ear infections, cavities, pneumonia, and food poisoning.

In some cases bacteria can cause damage by producing toxic chemicals, but most bacteria cause illness simply by their presence. When they enter the body, they replicate quickly and damage the cells and tissues living near these massive numbers of unwanted guests.

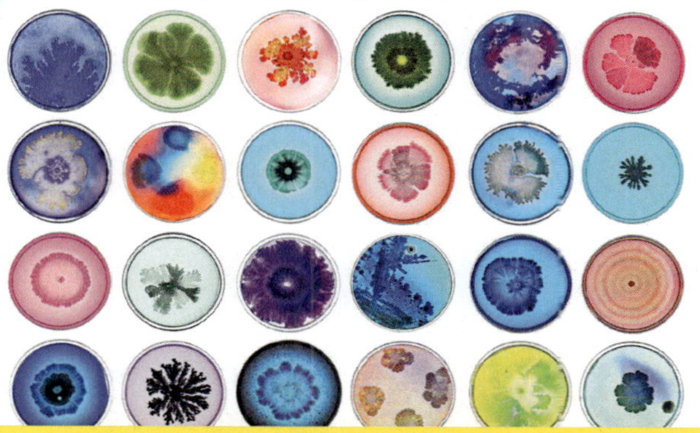

Scientists grow colonies of bacteria on petri dishes like these in order to study them. Some playful scientist had fun in their lab one day by making artwork out of these bacterial colonies.

Viruses

Unlike bacteria, **viruses** (vahy-ruh-ses) are not cells. They are made up of small particles of genetic material, like DNA, and are protected by a protein coat known as a capsule. Viruses make people sick by "hijacking" the cells of living organisms. With needlelike instruments, they inject their genetic material right into the cell. Once inside, they use the cell's own machinery to make many more of the same viruses (copies), which eventually destroys the cell.

While most viruses are destroyed before doing much damage, in cases where they succeed in proliferating, the numerous new viruses created then spread around the body where they can infect and destroy other cells, making people sick in the process. Both colds and the flu are caused by viruses that invade the respiratory system, and there are many other types of viruses capable of infecting people, plants, and animals.

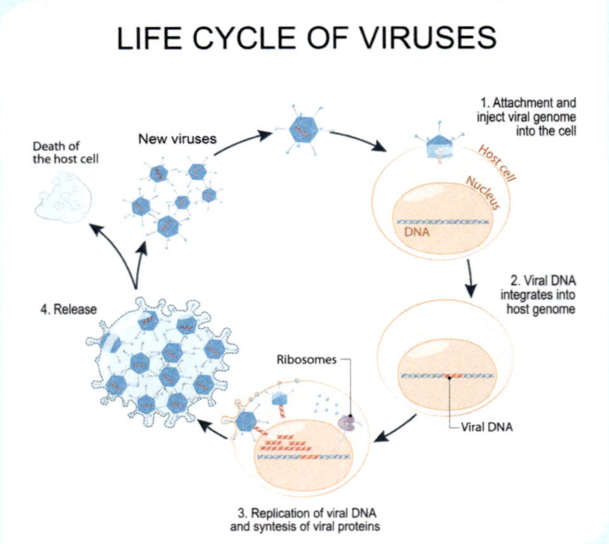

Because viruses are not cells, there is a controversy among scientists as to whether or not they are considered "alive." You might think it is simple to identify what is living and what is not, but what defines a living being has actually been debated by scientists for a long time. There are seven characteristics generally shared by all living beings—the ability to: respond to stimuli, grow over time, produce offspring, maintain a stable body temperature, metabolize energy, consist of one or more cells, and adapt to their environment. How well viruses meet many of these criteria is not clear. For example, on one hand viruses do replicate themselves, however, they only do so with the help of an outside cell's machinery, and they lack the ability to reproduce without outside help. So, are viruses alive? We'll have to leave them as a definite maybe.

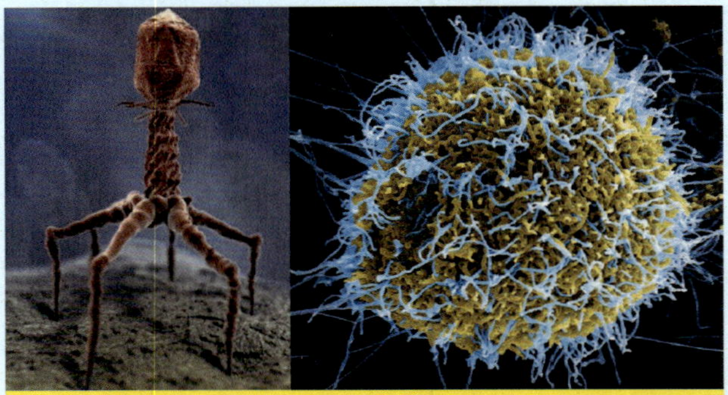

Actual pictures of the Escherichia virus T4 (left) and the Ebola virus attacking a cell (right).

Fungi and Protists

Fungi and protists are less common pathogens, but there are quite a few dangerous ones out there. Fungi include molds and yeasts and can cause diseases such as athlete's foot, ringworm, and thrush. Protists are single-celled organisms found all over. Some can cause diseases such as African sleeping sickness, malaria, and severe forms of diarrhea.

Wearing dirty socks is one of the ways people get athlete's foot, so remember to do your laundry!

Pathogens can make their way into humans in a variety of ways that include coming in contact with an infected person or animal; or an object such as food, water, or soil containing germs. Diseases caused by pathogens are referred to as infectious diseases since people suffering from those diseases can infect other people. (This is in contrast to noninfectious diseases, such as allergies, diabetes, and cancer, which cannot be transmitted from person to person.) To protect you from these little yet dangerous enemies, Hashem gave you three lines of defense.

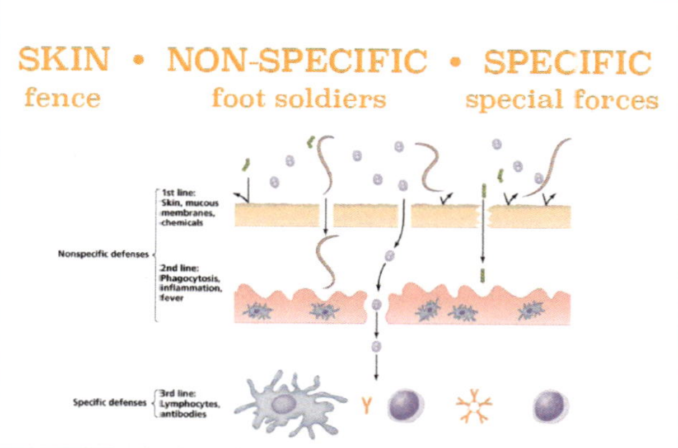

The three lines of defense against pathogens

First Line of Defense: Skin, Breathing Passages, and Stomach

The first line of defense consists of the skin, breathing passages, and the stomach. They collectively form a barrier or fence that prevents illegal entry to your body.

Your skin acts sort of like a barbed-wire fence that keeps out most intruders. When pathogens land on the skin, they are met by oil and sweat full of chemicals that are helpful to us because they are destructive to pathogens. In addition, new skin cells are constantly growing and old cells are constantly falling off, causing any pathogens that might have been resting on them to fall off with them. Even if pathogens manage to withstand the chemicals and don't fall off with old skin cells, a layer of tightly packed dead skin cells that is always present on the body prevents most pathogens from entering. If, however, you get cut, that gives pathogens a short window of time to infiltrate before a scab forms and the border is closed safely again.

This image of skin, taken by an electron microscope, shows the layer of "armor" provided by dead skin cells that prevents entry of pathogens.

Because we need to breathe and eat, our nose and mouth open and receive numerous uninvited pathogens, which must then be caught and removed. As we discussed earlier, breathing passages such as the nose, trachea, and bronchi are lined with mucus and cilia, which trap and remove most pathogens entering the respiratory system. In the event of a heavy invasion, irritation caused by pathogens triggers a powerful force of air to blow them out of our system. We call these sneezes and coughs—and that is why it's important to turn away when we sneeze and cover our mouths when we cough. Otherwise we are spraying pathogens on everyone around us.

Some pathogens are found inside of foods—those are met immediately by destructive chemicals in saliva when we take our first bite.

The ones the saliva doesn't stop soon meet up with the powerful acid in our stomach, which kills all but a few microbes. So, the vast majority of swallowed pathogens are destroyed by the saliva or stomach acid. The ones hardy enough to get further have to contend with the body's second line of defense, the inflammatory response.

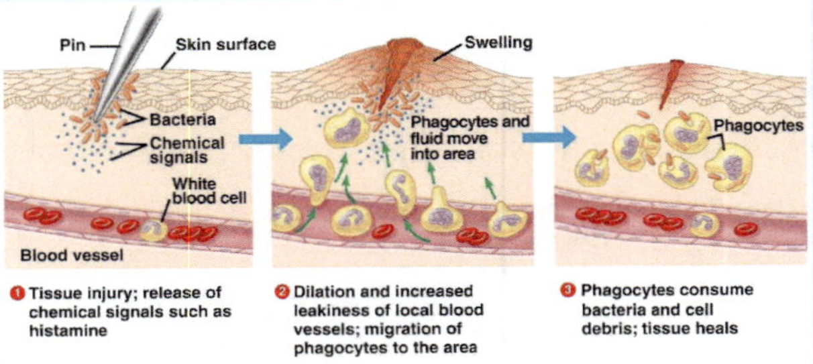

The inflammatory response

Second Line of Defense: The Inflammatory Response

If pathogens make it past the body's first line of defense, they begin to damage cells, which then release chemicals that send out messages that they are under attack and it's time to send in the troops! The **inflammatory response** causes fluid, carrying white blood cells from blood vessels, to flood into the tissues in the area being damaged. All white blood cells are disease fighters, but there are different types of them with different functions. The white blood cells that are sent in as part of the inflammatory response are called **phagocytes** (fag-uh-sahyts). They swallow and destroy pathogens using digestive enzymes that tear apart the invaders.

In order to bring more phagocytes to the scene during an inflammatory response, blood vessels widen in the area damaged by pathogens. The enlarged vessels allow more blood containing more phagocytes to be "deployed on the battlefield." Due to the enlarged blood vessels and fluid that leaks from them, the affected area becomes red and swollen or inflamed. This causes the area to feel warmer than usual, and can be felt by gently touching an area that is red from inflammation. People refer to this part of the body's defense system as an inflammatory response because the heat (or "flames") coming from the battle taking place inside of you can be felt.

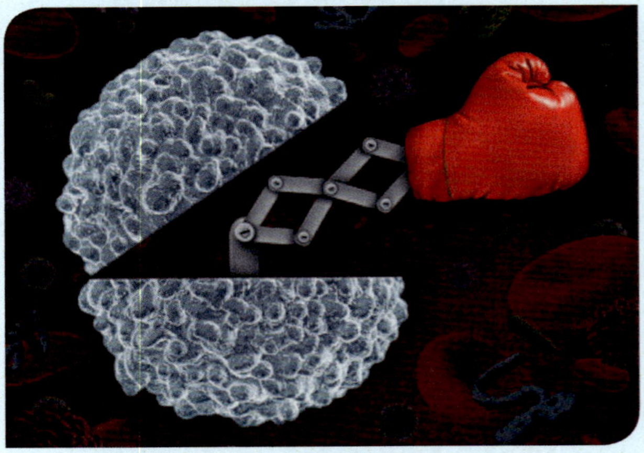

In some cases, the body needs to give its soldiers an advantage over the pathogens they are trying to fight off. This is accomplished by raising the entire body's temperature in what we call a fever. While fevers are certainly not fun, the response is designed to help you, since most pathogens do not grow well or reproduce quickly at high temperatures. The rise in heat prevents the rapid growth and spreading of pathogens and makes it easier for the white blood cells to ward off the attack.

Third Line of Defense: The Immune System

If an infection is bad enough to cause a fever, the third line of the body's defense forces are called up for duty. The third line of defense is called the **immune response**. It is controlled by the immune system and is different than the first two lines of defense. The skin, air, and food passages and the inflammatory response do not target any particular pathogen, but rather try to keep any foreign elements from entering the body. The immune response, however, is a specific disease-fighting system. Cells of the immune system can distinguish between different types of pathogens and respond with an attack targeted at that pathogen alone. At this point, the body is in big trouble—that's why it sends in its special forces to make sure we win the war.

The white blood cells that can differentiate between different kinds of pathogens are called **lymphocytes** (lim-fuh-sahyts) and there are two kinds: T-cells and B-cells. The T- and B-cells work together to combat an infection.

It is the job of the millions of different types of T-cells to identify the particular type of pathogen that has entered the body. Hashem placed thousands of receptors on their surface in order to do that. Each pathogen has unique marker molecules (antigens) and each is recognized by a T-cell that has a unique pattern of receptors that can identify the particular pathogen. When a T-cell recognizes antigens of a pathogen, it is triggered to rapidly multiply so more T-cells that can recognize that pathogen will be present.

The T-cells that directly attack the pathogens they were programed to destroy are referred to as killer T-cells. Other T-cells, called helper T-cells, signal and activate B-cells, which produce special proteins called **antibodies** (anti-bod-ees). Antibodies are perhaps the most powerful weapon in the immune system's arsenal of weaponry. Each type of antibody has a different structure that fits together with a particular antigen marker on the surface of the pathogen it was designed to destroy, like a lock and key. Like T-cells, each B-cell is unique and produces one type of antibody that can find and fight one type of antigen marker on a particular pathogen. While antibodies don't directly kill the pathogens, they do a number

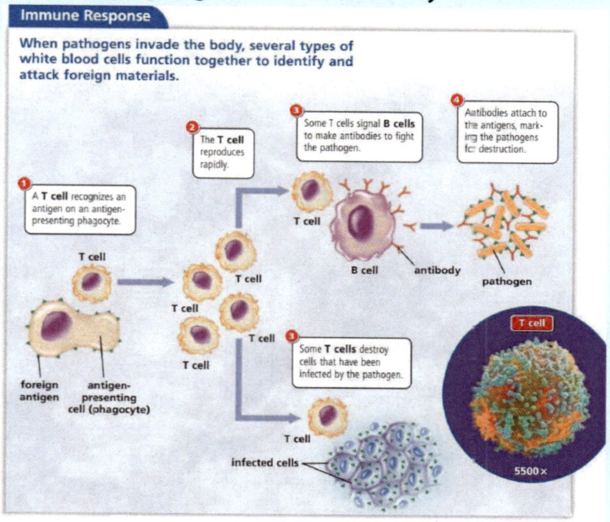

of things to ensure the pathogens will ultimately be eliminated.

Some antibodies work to destroy pathogens by marking them for destruction by other cells. Their presence makes it easier for phagocytes to recognize the pathogens as foreign materials and eliminate them. Other antibodies lock onto pathogens and cause them to clump together, which prevents them from spreading and makes them easier to destroy.

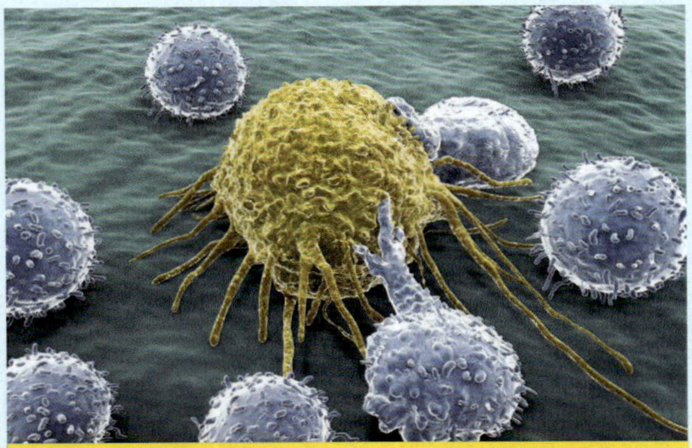

This scanning electron microscope image shows a group of T-cells moving in on a cancer cell to battle the microscopic beast.

In addition to actively fighting off infections, the immune response produces some T- and B-cells that remain in the body and "remember" how to most effectively fight the type of pathogen that caused the illness long after those pathogens have been destroyed. These "memory" cells ensure that if the pathogen ever tries to move into your body again, it will be met with a swift and powerful immune response that will eliminate it without you ever getting sick.

Immunity

The body's ability to destroy pathogens before they cause disease is called **immunity** (ih-myoo-ni-tee) and there are two types: active and passive immunity. **Active immunity** comes when a person's own immune system produces antibodies to a pathogen—for example, chicken pox antibodies created in the body of a person who had the chicken pox virus. If you ask most adults, they will tell you they had chicken pox once when they were young, it itched a lot and was very unpleasant, and have never had it again since. The reason for this is that after the body's immune response defeated the chicken pox virus, it left some T- and B-cells to circulate in the bloodstream that remember the antigen found on the chicken pox virus and know how to fight it. In the future, when that person is exposed to the chicken pox virus again, like when their own children get it, the memory cells start the immune response so fast that the person rarely gets sick.

A second way to acquire active immunity is by receiving a **vaccine** (vak-seen). A vaccine is a weakened form of the pathogen, or one that has already been killed. The either weakened or inactive (dead) pathogen no longer has the ability to harm us in a significant way, but can still trigger the immune response because it has the same antigens on its surface as the actual pathogen. The immune response against the vaccine creates T- and B-cells that remember the pathogen and will rapidly eliminate it if it enters your system in the future—even though you never actually endured that disease.

Vaccines are most often given by injection, but

some are given orally (through the mouth) or with nasal sprays. The process by which vaccines are introduced to a person's body to produce immunity is referred to as vaccination or immunization. It has saved millions, maybe even billions, of people from getting potentially life-threatening diseases such as polio, tetanus, and diphtheria.

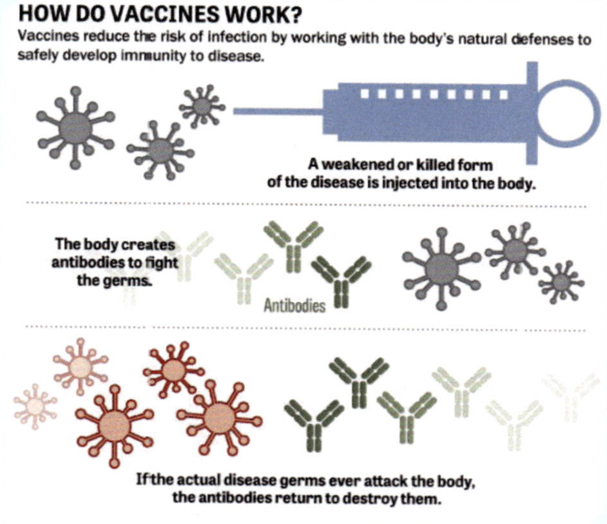

Some diseases are very rare and are therefore not routinely vaccinated against. These include rabies, smallpox, and yellow fever. In the event that someone was infected by such a disease, they can be given **passive immunity** by injecting antibodies from a source other than the patient's own body. While the antibodies are usually successful in saving the person's life, passive immunity does not last long and the person will become susceptible to infection again in a few months' time. Passive immunity to certain diseases is also passed from a mother to her baby before it is born. It is a great gift, since a newborn would otherwise not have an adequate immune system to fight off disease. After birth, the baby carries the immunity for a few months, which gives the child's own immune system a chance to strengthen until it can better fight off pathogens on its own.

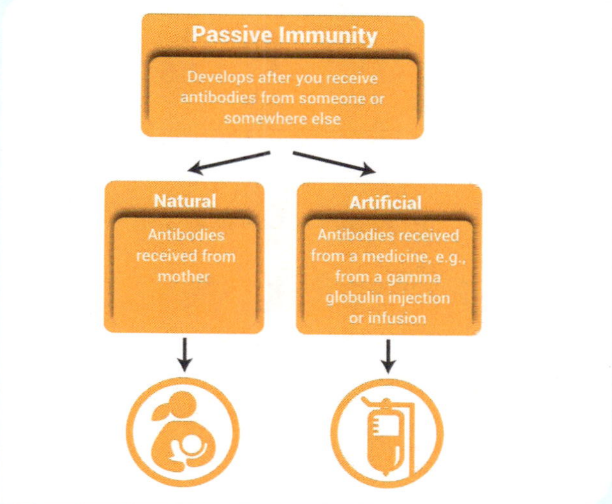

Allergies

In some cases, the immune system can have an abnormal reaction to a typically harmless substance. That abnormal reaction is called an **allergy** (al-er-jee). Substances which cause allergic reactions are known as **allergens**. Allergens include some foods such as peanuts, dust, plant pollen, pet dander (flecks of skin that fall from cats, dogs, and other furry creatures), or even certain medicines. When a person is allergic to something, the immune system mistakenly believes that this substance is harming the body and tries to fight it off as it would a real pathogen.

In an attempt to protect the body from the allergen, the immune system makes antibodies. These antibodies, unlike the ones made during the typical immune response, signal cells in the body to release chemicals into the bloodstream to defend against the allergen "invader." It is the release of these chemicals that causes the allergic reactions.

Some common allergens

Reactions can affect the eyes, nose, throat, lungs, skin, and gastrointestinal tract. This causes symptoms such as rashes, coughing, sneezing, and breathing difficulties. Depending on the severity of the allergy, those symptoms can range from mildly annoying to very serious or even life-threatening. Future exposure to that same allergen will trigger the allergic response again.

Asthma is a condition that is sometimes caused by allergies. Asthma causes wheezing and a shortness of breath due to narrowing of the respiratory passages. It is most often treated with an inhaler that contains medicine able to reverse the effects and restore breathing to normal.

One chemical that is released, **histamine** (his-tuh-meen), is responsible for many of the classic symptoms of an allergy, such as sneezing and watery eyes. Some drugs can lessen the actions of histamines; they are called antihistamines. Nevertheless, often the best strategy is to try to avoid the allergen in the first place. While allergies are certainly not fun, it is a great blessing from Hashem that He gave us the wisdom to develop drugs that can relieve the symptoms of allergies and other illnesses.

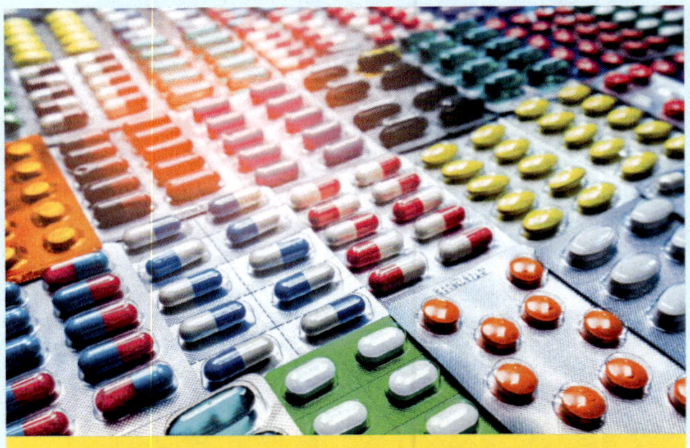

Remember to thank Hashem for every pill you take!

HEALTHY HABITS:
Immune Boosters

Can the immune system be improved? Does it even need improving? On the whole, our immune system does a remarkable job of defending us against disease-causing microorganisms. But sometimes it fails: A germ invades successfully and you get sick. Is it possible to fight back by boosting (strengthening) your immune system? There are numerous products available that have been proven to boost the immune system. Here are five of the most popular options:

Echinacea is an herb well known for its immune-boosting properties. It should be used at the very first signs of a cold, such as a sore throat or sneezing. Echinacea can only be used for 5 to 7 days because the body gets used to it and then does not respond anymore. It is useful for a short-term immune boost.

Elderberry is a bush that produces small, dark berries that are ready to harvest in the fall. It, too, is well known for its immune-boosting properties, especially its remarkable ability to prevent flu and cold viruses from spreading rapidly in the body. Unlike echinacea, elderberry can be taken for longer periods of time and many use it all winter long to avoid getting colds in chilly weather.

Honey. Raw, unpasteurized, unfiltered honey possesses antibacterial, antiseptic, anti-fungal properties. One tablespoon of honey can be diluted in 8 ounces (227 grams) of warm water with the juice of half a lemon and/or 1 tablespoon of raw, unfiltered apple cider vinegar, as an immune-boosting tonic to be used at the first signs of a cold.

HEALTHY HABITS -CONTINUED

Garlic. When raw garlic is crushed or chewed, its cells release a natural antibiotic and anti-fungal compound called allicin, which helps fight off infections. Garlic also provides the body with much-needed sulfur, which helps boost the immune system. In addition, raw garlic is rich in vitamins C, B1, B2, and B6; magnesium; zinc; and selenium; and also contains lower levels of other vitamins and minerals (B3, B5, calcium, iron, phosphorus, and potassium).

Chicken soup. As unbelievable as it may sound, chicken soup's ability to relieve the symptoms of a cold and speed up recovery is not just a grandma's tale. Research has proven that hot chicken soup increases nasal mucus velocity (helpful for getting out bacteria faster) and functions as an anti-inflammatory, making it a great short-term immune booster.

SECTION 3.3 REVIEW

1. What are pathogens and what are some examples of them?

____

2. What is the first line of defense your body has against pathogens?

____

3. Describe the inflammatory response.

____

4. Which type of white blood cells are used in the immune response and what is unique about them?

____

5. What is a vaccine and how does it provide immunity?

____

USE YOUR MIND

There are many rare diseases that could be prevented by giving vaccinations against them as part of the standard vaccinations given to children at birth. Why don't doctors recommend giving every possible vaccine, if doing so could protect people from most of the pathogens on the planet?

____

____

SECTION 3.4: EXPAND YOUR MIND

SEEING THE PLAN AND THE PURPOSE

We covered a lot of material in this chapter. It's tricky to absorb it all, while also keeping in mind the absolute miracles that are taking place within you at every moment. Just think about the fist-sized super pump sitting right in the middle of your chest—which has a built-in natural pacemaker that speeds it up and slows it down as needed, beats billions of times and yet never takes a break, and pushes a revolving cycle of blood that is equivalent to 2,000 gallons' (7,500 liters') worth through the 100,000 miles (161,000 kilometers) of your blood vessels every day.

Now think about those blood vessels that were designed with three different layers: smooth on the inside to allow easy flow, hard on the outside for protection, and with muscles in the middle so they can expand and contract according to the body's need for blood.

Now think about the superfluid we call blood that's flowing through your body. It's literally a river of nutrients carrying thousands of different substances that supply the 100 trillion cells of your body with everything they need. Not only does the blood transport the materials, it actually reworks and modifies many substances on their way to their destinations. Talk about a great delivery service! Rabbi Avigdor Miller, *zt"l*, compared blood's activities to a bakery truck that while en route to its customers is not only baking the cakes, but also grinding the wheat, mixing it with water, and making the flour too!

Miraculously, as these thousands of different nutrients are carried along in the blood, they don't interfere with each other. How could that be? We can't put two different medicines in one bottle because of the risk of an unexpected reaction. But in our blood, not only don't the thousands of nutrients interfere with one another, the blood brings each and every nutrient to exactly the right place. This is equivalent to a deliveryman who needs to make deliveries to a whopping 10,000 addresses.

Beyond that, the blood performs a garbage-collection process, carrying away the waste materials from the cells. And—much of the cellular waste products are carried away for elimination in the same blood flow that is carrying all the nutrients!

To illustrate what is happening inside blood, imagine a big pipe carrying milk from Manhattan to Brooklyn. At the same time, this pipe is carrying oil for furnaces and additional oil for cooking. There are all kinds of needed liquids in the same pipe, yet they remain clean and distinct. Meanwhile, that same pipe is bringing sewage from Brooklyn to Manhattan! Although the sewage is traveling in the same pipe, one does not contaminate the other. How all this happens inside of blood is truly a miracle of miracles.

In addition, blood carries a host of warriors that are on guard 24 hours a day without rest to protect the body from pathogens. These white blood cell guards prevent infiltration by swallowing up and thereby eliminating the intruders. In the event that the enemy does make it past those soldiers, special forces units are called up to the battlefield. They perform precise attacks against particular pathogens by using T-cells and by shooting missiles called antibodies that have sophisticated targeting systems that avoid everything except their intended target.

These concepts are actually only some of the very basics of what blood is all about. The miracles of blood could fill libraries. And so, when we see the blood we can say, "How abundant are Your works, Hashem" (*Tehillim* 104:24). Blood is so essential to human and animal life that the Torah states, "For the blood, it is the life" (*Devarim* 12:23). Life is maintained because of the miracles that take place in the blood—and it is truly a great kindness from Hashem.

(Based on *Ohr Avigdor*, *Chovos HaLevavos*, *Shaar HaBechinah*, pp. 191–192, by Rabbi Avigdor Miller)

SECTION 3.4: EXPLORE CHAZAL

TORAS EMES

Seeing that we have just learned about the circulatory system, I think it's appropriate to mention an idea by Rabbi Meir Leibush ben Yechiel Michel Wisser (1809–1879)—a Ukrainian scholar better known (by his initials) as the Malbim. He explained how the *Shulchan* (Table) and the Menorah in the Mishkan represent the heart and lungs of the Jewish people. Before we explain the details, a brief explanation of the connection between the body and the Mishkan is in order.

The Malbim explains in *Parshas Terumah* that the Mishkan was designed for a dual purpose: to serve as a vehicle where the Shechinah (Divine Presence) could dwell among our nation, and to enable the Jews to live with a tangible connection to Hashem.

Hashem designed the world in such a way that perfectly righteous men could influence creation and bring great blessings and bounty into the world. However, understanding that there were not many individuals capable of achieving such perfection, Hashem gave the Jewish people a place that could substitute for that "perfect man" and be a channel for bringing an abundance of blessings into the world. Although each individual may not have attained perfection in every one of their character traits and good deeds, together the strengths of each Jew would unite in the Mishkan to form a collective "perfect man."

The Malbim goes on to explain which part of the Mishkan corresponded to which part of the "perfect man's" body. The *Shulchan* represented the heart and the Menorah represented the lungs. Just like in a body, the chest is below the head, so too the *Shulchan* and the Menorah were situated below the *Aron* (Ark), which represented the mind. As such, they were located in what would be the chest cavity of a person. The chest is the place where the heart and lungs, which bring life to the rest of his body, are located.

The *Shulchan* represented the heart. Just like the heart nourishes, gives strength, and carries life to the whole body by way of the blood, so too would the whole world receive its nourishment and sustenance by way of the Table in the Mishkan. To accomplish this, Hashem instructed that the *Lechem HaPanim* (showbread) be placed on the *Shulchan* so that from the bread of Hashem, blessing and sustenance would pour onto the whole world.

The Menorah represented the light of the intellect and knowledge. While the *Aron* represented Torah knowledge and prophecy, which come directly from Hashem, the Menorah represented a person's own knowledge and understanding that come from working diligently to gain deep clarity and understanding. It had branches (seven of them), representing the branches of the lungs, which bring pure air, carrying oxygen, to the brain. When those branches are lit, they represent the seven branches of wisdom that can be acquired. Those branches of wisdom are found in: the elements of the world; the human race (as a whole); man's physical and spiritual traits; the species of other living creatures; plants and minerals; the sciences, crafts, and industries; and above all, the wisdom found in the Torah and its laws. The six branches of the Menorah all point toward the center flame, which represents the wisdom found in the Torah, since the wisdom of Hashem is the highest and most important of them all.

This is only one of many examples where the Malbim's extensive knowledge of science and nature enabled him to have deep and profound insights into Torah.

EXPERIMENT:

HEARTBEATS

Corresponds to Section 3.1

Materials:

- Mini marshmallows
- Toothpicks
- Ruler
- Plastic Straws

Procedure:

1. Working with a partner or groups of three, make sure each person has one mini marshmallow.
2. Make narrow four-inch sticks by cutting the plastic straw lengthwise into strips or sticks so that there is one stick for each marshmallow.
3. Insert the stick into the marshmallow.
4. Lay your left hand, palm up and still, on the table.
5. Find your wrist pulse (where you feel the pulsing sensation).
6. Stand the marshmallow on top of the pulse point and carefully watch the tip of the straw.
7. Have one partner count the number of times the stick twitches in 15 seconds and then multiply that number by four to get the pulse rate (The average adult pulse rate is between 65-75).
8. Repeat two or three times to get each person's resting heart rate (pulse rate while not doing strenuous activity.

9. Now jump up and down or run in place for one minute and recount the pulse rate.

Questions:

1. What was your resting heart rate?____

2. What was your heart rate after doing strenuous activity?___________________________

3. Was your resting heart rate higher or lower than the average adult heart rate? Why do you think this is?____

Fun Facts: Here is a list of the resting heartrates per minute of some animals: Bat 750, Camel 30, Cat 120, Chick 380, Chicken (Adult) 280, Cow 64, Dog 110, Elephant 35, Giraffe 65, Goat 90, Guinea Pig 280, Hamster 450, Horse 45, Lion 40, Monkey 190 ,Mouse 520, Rabbit 205, Sheep 75, Skunk 166, Squirrel 250

EXPERIMENT:

A LITTLE BIT OF BLOOD

Corresponds to Section 3.2

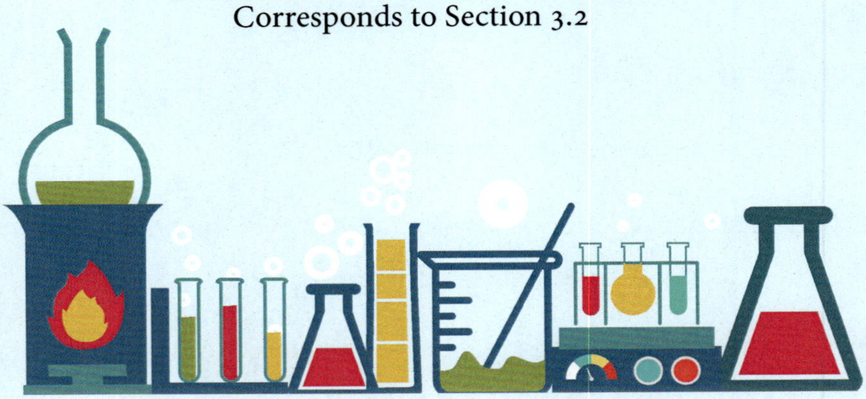

Materials:

- Plastic water bottle (1.5 liters)
- Yellow food coloring
- Salt
- Cheerios
- Mini marshmallows
- Purple Pompoms
- Newspaper

Procedure:

Put newspaper down on the table you are working on.

Fill up the plastic bottle half way up with water and add two drops of yellow food coloring (representing blood plasma) and a pinch of salt (representing minerals and chemicals in blood).

One person in class (can be teacher) takes a 1-quart Ziploc bag and adds a whole bottle of red food coloring and then fills it up with cheerios (representing the blood cells) mixing them up until they turn red.

Each student or group should add a handful or more of the cheerios if available to their bottle.

Add mini marshmallows (representing white blood cells) and purple pompoms (representing platelets) to the bottle (both less plentiful than the red blood cells).

Section 3.4: Exp: Expand, Explore, Experience

Essay: Describe the importance of each of the components in your blood model.

EXPERIMENT:

READY, SET, INFECT!

Corresponds to Section 3.3

Introduction

In this experiment you will be recreating the epic fight between pathogens and the immune system in your classroom. Anyone who has seen live footage of a human white blood cell chasing and capturing a bacterium in the body would think he or she was witnessing a game of cops and robbers. The white blood cell determinedly chases the invader until the cell reaches its goal of ingesting and then "spitting out" the villain. Unfortunately, the excitement of the chase does not easily translate onto the written page of a science textbook.

Ready, Set, Infect! is a role-playing game that allows students to experience how dynamic the human immune system must be in order to protect the body against infection via its three primary lines of defense: skin, saliva, and stomach acid; white blood cells (through inflammation); and antibodies (via the immune response).

In this exercise, you will actively engage in and develop a model of how each component interacts with the other to protect the body from invasion and illness.

Review

The immune system relies on three main lines of defenses, two of which are nonspecific to the pathogen, and one of which is specific to fighting a particular type of germ.

Nonspecific Defenses

- Skin is an effective barrier to the entry of microorganisms, while cilia, tears, sweat, saliva, mucus, and earwax trap or wash microorganisms away.
- The inflammatory response (white blood cells). Various types of white blood cells either surround and engulf microorganisms or kill their targets by releasing damaging chemicals. For

example, histamines promote the dilation of local blood vessels and the release of more white blood cells from the vessels that will enter the infected tissue and help it heal.

Specific Defenses

- The immune response (antibodies). A type of white blood cell known as B-cells produce antibodies against foreign antigens, while additional white blood cells (T-cells) enhance the immune response and kill foreign cells directly.

The game you are about to play is a competition between two teams. Students on one team represent an invading source of infection (antigens), and students from the other team are members of the human immune system (skin, white blood cells, and antibodies).

Materials

- antigen-antibody cards (printed out from *Teachers' Resources*)

Procedure

1. Before the lesson starts, the antigen-antibody cards should be cut along the dotted lines to separate the two roles of antigens and antibodies. There are enough role cards for a class of 28 students, but extra cards can be copied for larger classes. The role cards can be enlarged and/or laminated for ease of use.

2. Store the desks and chairs at the side of the classroom and clear enough space for students to walk quickly around the room. To ensure safety, no running is permitted during the activity. Tagging by touch must be gentle—no pushing.

3. Each student can pick out a card with their eyes closed or the teacher may hand them out. It's important that students do not know who their matching antigen or antibody is before the start of the game. The cards will determine the roles of the participants as follows:

    a) Skin. This role is taken by the teacher. The teacher stands at the inside of the classroom door and is in charge of keeping it closed. The door remains unlocked, but the teacher must keep it from being opened by members of the antigen team that is standing outside the room.

    b) White blood cells (only phagocytes). These students are in charge of standing inside the classroom, by the door, to prevent invading antigens from proceeding further into the room. The phagocytes must "engulf" antigens by gently placing both hands on an antigen's shoulders. If only one hand is touching the antigen, then it can escape and roam free. If the antigens successfully make it past the white blood cells, then the white blood cells summon the third line of defense: the antibodies, who have been waiting for that stage to begin.

c) Antibodies. These students are antigen specific and should wait further inside the classroom. Their cards are uniquely shaped; the antibodies are only allowed to capture the antigen who has the card with the shape that fits into theirs. (Remember our lock-and-key analogy?) The point of this stage is for each antigen to find the right antibody. This is why it's important that students do not know who their match is before the start of the game—that would spoil the fun!

d) Antigens. These students remain outside the classroom door at the start of the game. Once the teacher announces that a cut or a breach in the body has occurred, the teacher (playing his or her role as the skin) opens the door widely and lets the antigens attempt to enter the classroom.

4. If the antigens are successful in getting past the skin (the first line of defense), then they must battle the white blood cells (the second line), who will try to tag the antigens by touching the antigens' shoulders. If the antigens are successful in escaping the white blood cells, then it's up to the antibodies (the third line of defense) to tag the antigens. Once an antigen is tagged, it must present its card to the antibody. If the cards are a match, then the antibody and antigen sit down on the side of the classroom next to one another. If the cards do not match, the antibody must release the antigen and attempt to capture the next antigen until it finds and captures its match.

The teacher will act as the narrator, timekeeper, and referee of this game. As the narrator, the teacher will report on the action in real time (e.g., noting when the antigens pass each line of defense or how many antigens are still at large).

Note: Although B-cells have an important role in making and proliferating antibodies, their role has not been included in this version of the game—in order to give as many students as possible an active part in the chase as antibodies or antigens.

5. Play the game as described for about 4 to 5 minutes. If there is a single surviving antigen at the end of the game, then the class is infected with the disease written on the antigen's card and the body will become ill (e.g., catch a cold) as the antigen multiplies. If all the antigens are defeated, then the immune system has successfully kept the body healthy.

6. Repeat the game at least twice, with the teams changing roles so that each student has a turn to be an antigen or an antibody at least once.

If the body becomes infected the first time the game is played, then before the next round, the teacher will assign more students the role of white blood cells and antibodies (with less students being antigens), since the body's immune system will already be hard at work producing more defenses.

7 After the game, students should summarize the process by which the immune system defends the body against pathogens.

8. Draw a model of the immune system (it can be a concept map, a comic strip, or any other form of expression that is appropriate). Include terms from each of the three lines of defense in your diagram.

Sample Concept Map

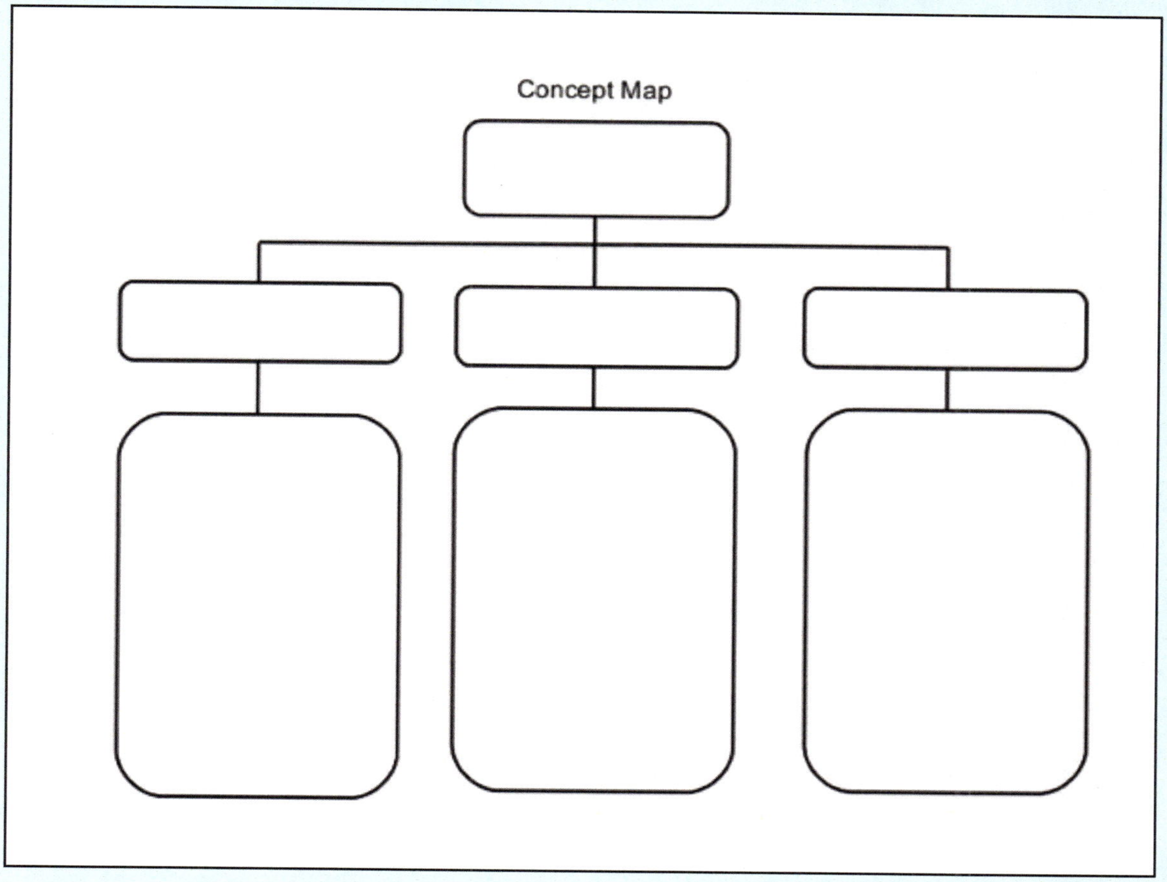

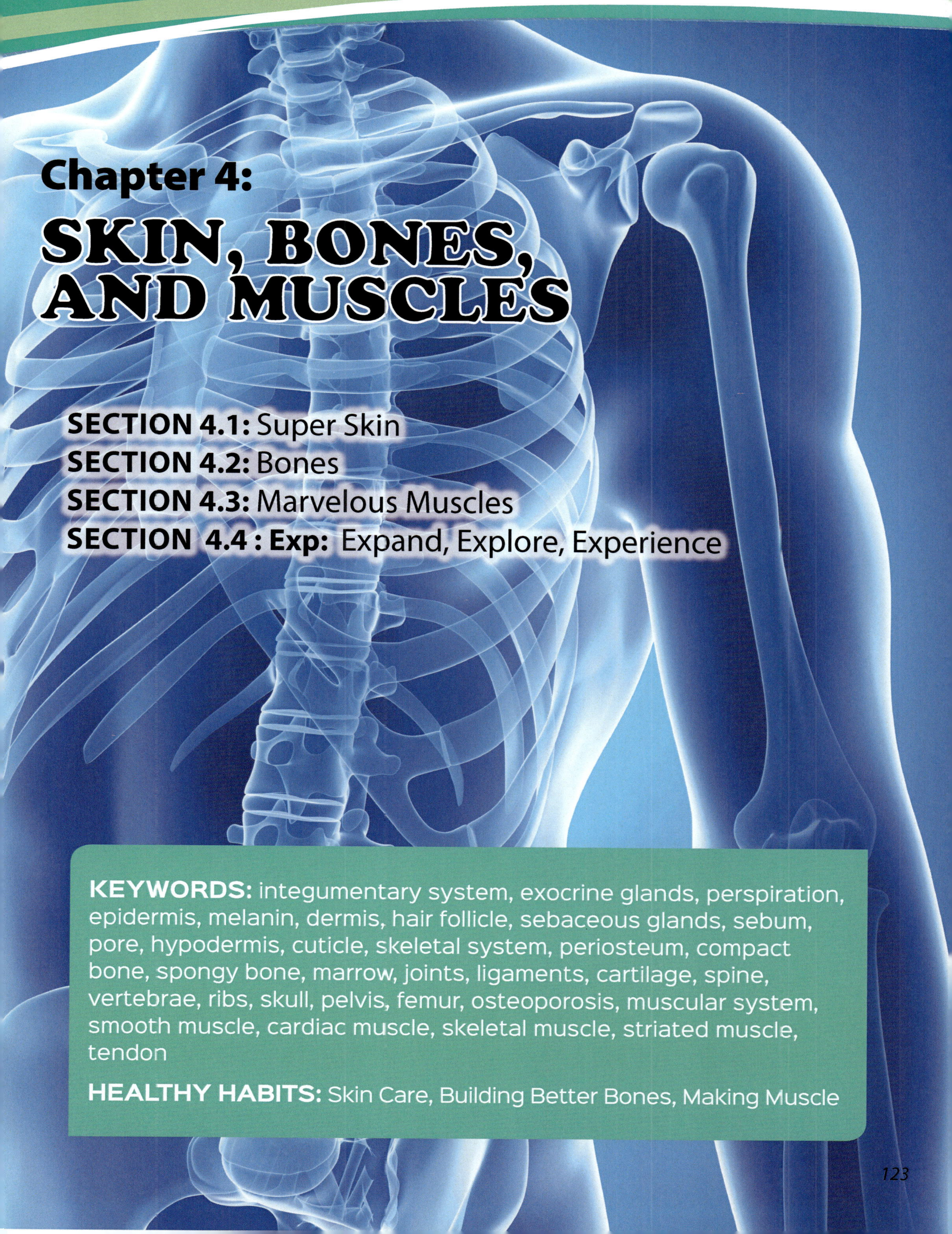

Chapter 4:

SKIN, BONES, AND MUSCLES

SECTION 4.1: Super Skin
SECTION 4.2: Bones
SECTION 4.3: Marvelous Muscles
SECTION 4.4 : Exp: Expand, Explore, Experience

KEYWORDS: integumentary system, exocrine glands, perspiration, epidermis, melanin, dermis, hair follicle, sebaceous glands, sebum, pore, hypodermis, cuticle, skeletal system, periosteum, compact bone, spongy bone, marrow, joints, ligaments, cartilage, spine, vertebrae, ribs, skull, pelvis, femur, osteoporosis, muscular system, smooth muscle, cardiac muscle, skeletal muscle, striated muscle, tendon

HEALTHY HABITS: Skin Care, Building Better Bones, Making Muscle

SECTION 4.1: SUPER SKIN

Imagine there was a suit you could put on that could cool you off in the heat, warm you up in the cold, and even repair itself when it gets torn. How much do you think these would go for in a store? The reality is that your skin does much more than that. It is a true "super suit" designed by Hashem, with many extraordinary properties made to "suit" your body's needs.

Hashem is to thank for creating skin with super powers!

Skin is the largest part of the **integumentary** (in-teg-yuh-men-tuh-ree) **system**, which includes all elements of the external covering of the body, including skin, hair, nails, and sweat glands. In addition to being the largest element of the integumentary system, skin is actually the largest organ in the body. It performs many important functions, including: protecting the body, regulating its internal temperature, sensory reception, eliminating wastes, and vitamin D production. Let's explore these fantastic functions of your skin.

Protection

One of the main functions of skin is its role in protecting the body. As discussed in chapter 3, skin protects the body from pathogens and other harmful substances by forming a barrier that prevents their entry. In addition to keeping substances out, it has an equally important, if not more important, role of keeping substances in. The barrier provided by skin prevents the loss of important fluids such as water.

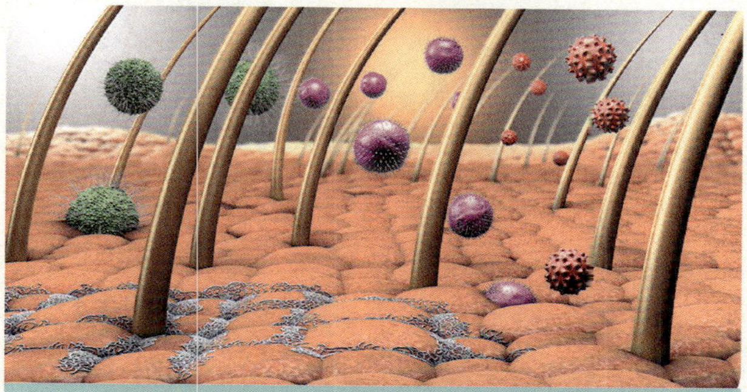

Skin keeps you safe from a host of pathogens that want to make the inside of your body their new home.

Temperature Regulation

Another function of skin is how it regulates the internal temperature of the body. While it may not be too pleasant, you can survive in a large range of outdoor temperatures ranging from the blistering heat of the summer to the bitter cold of the winter. Your insides, however, are far more sensitive and cannot go far off the optimal range of normal temperature, which is 97.7 to 99.5 degrees Fahrenheit (36.5 to 37.5 degrees Celsius). In order to accomplish this heat regulation, Hashem designed the skin with a few features that can raise the internal body temperature to accommodate cold weather and lower the body temperature to accommodate the heat.

The 2.6 million sweat glands in your body help cool you down in the heat.

The most effective way of cooling your body down in the heat is accomplished through sweating. Sweat is produced from sweat glands found all over the skin. Some glands secrete materials inside the body and are called endocrine glands (more on that in chapter 5) and others, such as the sweat glands, secrete substances outside the body and are called **exocrine** (ek-suh-krin) **glands**. When the outside temperature is high, sweat glands release bodily fluids combined with salt to keep the body temperature from getting too high.

Blood Vessel Dilation (Expansion) and Constriction

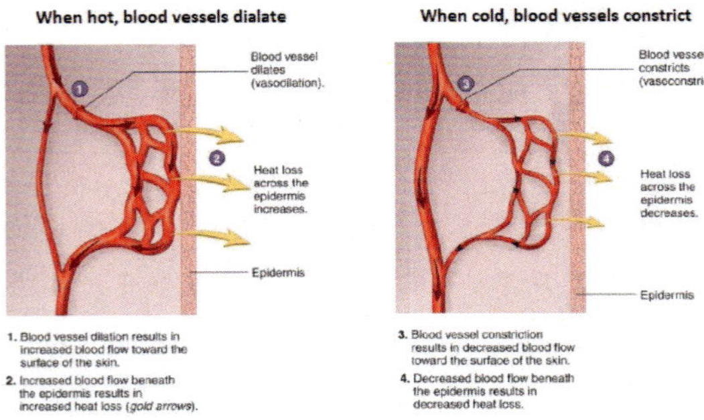

Temperature receptors in the skin tell the glands when to release the sweat. These receptors send signals to the body, stimulating sweating in hot weather. When the temperature outside is high, the many blood vessels that run through the skin enlarge and the amount of blood that flows through them increases. This increase in blood flow triggers sweat glands to release a fluid made up mostly of water and salt, known as **perspiration** (pur-spuh-rey-shuhn). Perspiration carries heat away with it—from inside your body to the surface of the skin where it settles. When perspiration comes in contact with the air, it evaporates (meaning it changes from a liquid to a vapor), and that heat is carried away by the wind.

Sometimes your skin saves you from the cold!

In addition to keeping you cool, skin can warm you up as well. There is warmth in the blood, which is why your body generally feels warm, but this heat is constantly being lost to the surrounding environment to a certain degree. When blood flows closer to the surface of the skin, more heat is lost. When skin receptors detect there is a drop in temperature (like when it's cold outside), the blood vessels constrict (get smaller), causing less blood to flow close to the surface. This prevents heat from being lost and keeps your internal body warmer.

When the body's internal temperature drops very low, skin receptors send signals to the body to start shaking. The rapid movement generates heat that can warm you up in the bitter cold. Yes, when you are shivering from cold, you are actually involuntarily helping yourself warm up a little. (But it's still a good idea to go inside and really warm up.)

Sensory Reception

The skin plays a vital role in virtually everything we do by sensing what is going on in the environment around us. The skin can sense pressure, pain, and temperature when it comes in contact with other people or things. The perception of pain is of particular importance, as it lets you know when something may have injured the body or is at risk of being harmed, for example by mistakenly touching s a hot pot on the stove.

Pain often protects us from getting seriously injured.

It is easy to overlook the gift of sensory reception, but observing the massive and sometimes life-threatening difficulties of those who have lost it can remind us how important it is. About one in every one million people are born without a sense of pain. People with this unfortunate disorder cannot feel hot objects, or cuts and scratches on their skin, or insect bites. Those with this disorder often suffer from severe self-inflicted injuries from an early age, which can lead to premature death. While we certainly don't enjoy pain, it is a vital built-in "fire alarm" that aids in our survival, and it is truly a great kindness from Hashem.

Eliminating Wastes and Producing Vitamin D

Two other functions of the skin are waste elimination and vitamin D production. Perspiration excreted by sweat glands contains dissolved materials that come from the breakdown of chemicals during cellular processes. So when you perspire, your skin is helping to eliminate wastes from your body.

A little bit of sunshine is all you need to make vitamin D.

Lastly, some skin cells are designed to produce vitamin D—but only when they are in the presence of sunlight. Vitamin D helps our digestive system absorb the calcium found in our food. Calcium is essential for healthy bones, so the skin is actually helping us have healthy bones. With just a few minutes of sunlight, the skin can produce all of the vitamin D the body needs for the day.

Epidermis

The skin is made up of three layers, called the epidermis, dermis, and hypodermis. Hashem designed each layer with its own important role in carrying out the functions of skin. The outermost layer is called the **epidermis** (eh-pih-dur-mis). Even though it doesn't seem like anything is happening when you look at your

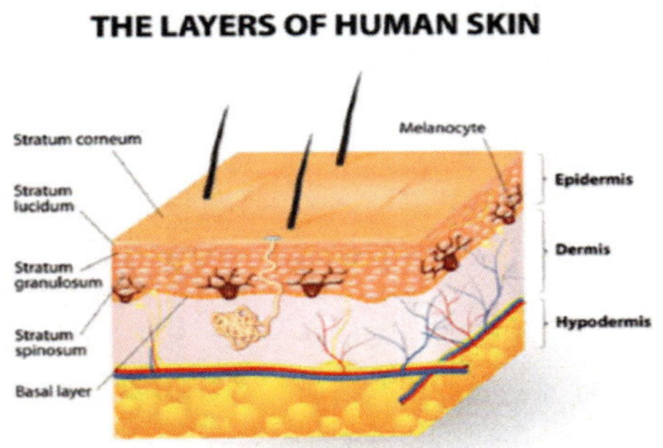

The three layers of skin

At the bottom of the epidermis, just under the surface of the skin, new skin cells are forming all the time. When the cells are ready, they start moving toward the top of your epidermis. This trip takes about two weeks to a month. As newer cells continue to move up, older cells near the top die and rise to the surface of your skin. What you see on your hands (and everywhere else on your body) are actually dead skin cells. The dead cells are still tough and strong, just right for covering your body and protecting it. But they only remain where you can see them for a little while before flaking off. Every minute of the day we lose 30,000 to 40,000 dead skin cells off the surface of our skin—even if we don't suffer from dry skin.

Most of the cells in our epidermis (95%) work to make new skin cells, and the other 5% produce a substance called **melanin** (mel-uh-nin). Melanin is what gives skin its color. The darker our skin is, the more melanin we have. When we go out into the sun, these cells make extra melanin to protect us from getting burned by the sun's ultraviolet, or UV, rays. The more time we spend outside in the sun, the more melanin our body produces, and therefore, the darker our skin becomes. That's called getting a tan. And, unlike the melanin we are born with, this extra melanin eventually fades because the tanned skin cells get replaced by new cells that do not have the extra melanin. So the old color of your skin returns if you stay out of the sun.

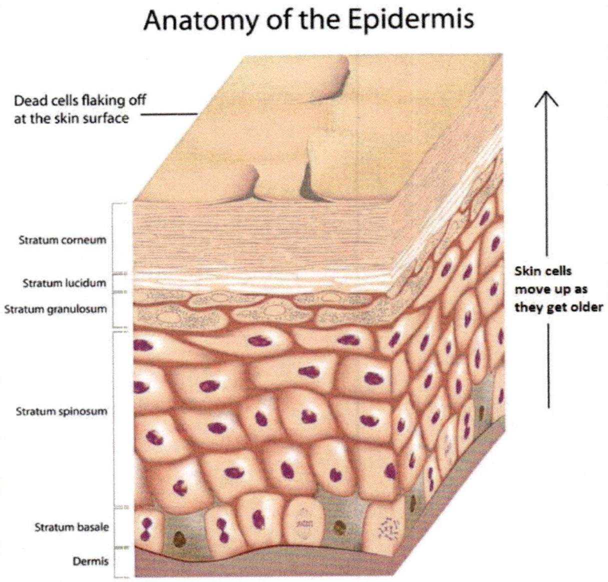

The epidermis is made up of layers that change as more cells move up to the surface.

Dermis

The next layer down is the **dermis** (dur-mis). We can't see our dermis because it's hidden under our epidermis. The dermis contains nerve endings, blood vessels, roots of hairs, oil glands, and sweat glands. It also contains collagen, which makes the skin tougher, and elastin, which makes the skin more flexible. The nerve endings in our dermis tell us how things feel when we touch them. They work with our brain and nervous system, so that our brain gets the right message about what we're coming in contact with.

SKIN ANATOMY

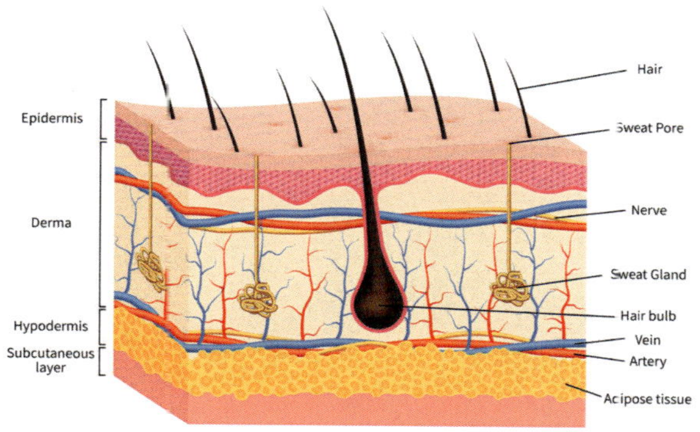

The dermis is also full of tiny blood vessels. These keep our skin cells healthy by bringing them the oxygen and nutrients they need and by taking away cellular waste products. As the dermis gets older, it gets thinner and easier to see through, which is why it is often easier to see the blood vessels in older people than in children.

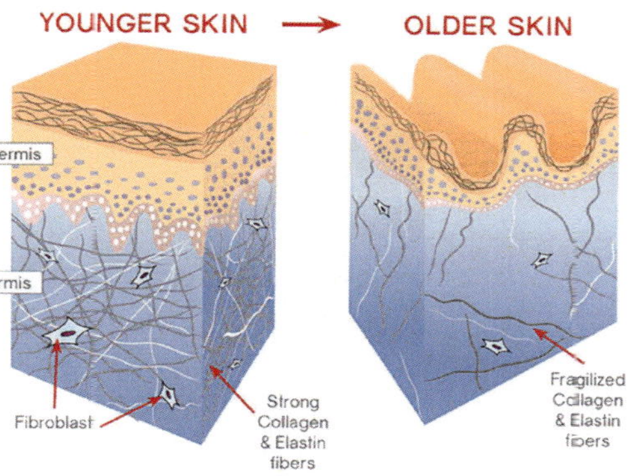

This layer of skin is also where hairs grow from. Each hair on the body grows out of a tiny tube in the skin called a **hair follicle** (fahl-ih-kul). There are hair follicles all over the body, except on the lips, the palms of our hands, and the soles of our feet. This may seem strange, as you don't associate something like your forehead as being hairy, but there are tiny hairs that are clear, so they generally go unnoticed.

Properties of Sebum

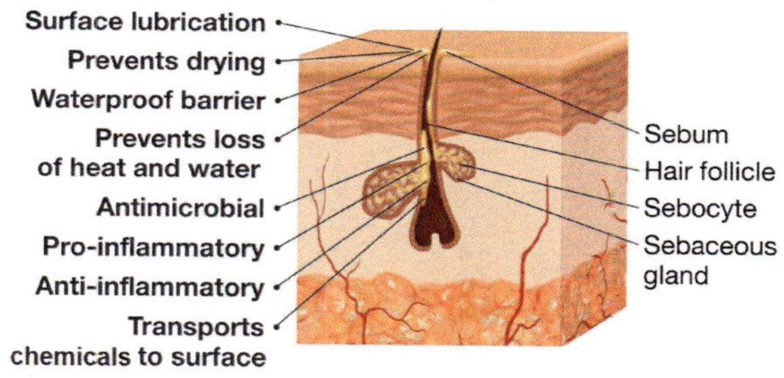

Another feature of the dermis are the tiny oil glands found there. They are called **sebaceous** (sih-bay-shus) **glands**. They are connected to each hair follicle and are always producing sebum. **Sebum** is our skin's own natural oil. It rises to the surface of the epidermis to keep the skin lubricated and protected. It also makes our skin waterproof. If you spend enough time in the water, though, the sebum comes off and water gets in, making your fingers look wrinkled.

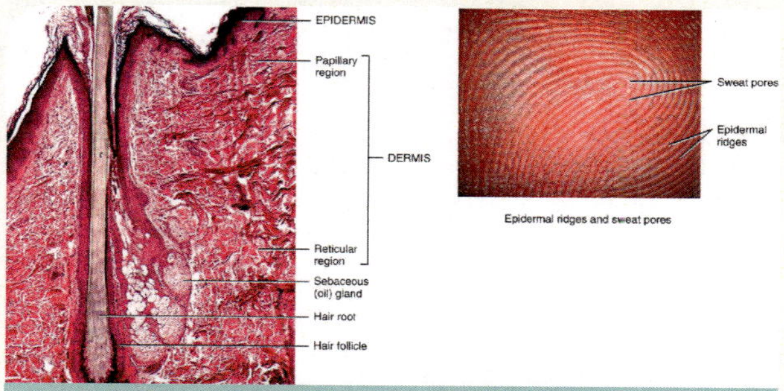

Actual picture of skin under a microscope and the location of sweat pores

Sweat also comes from the dermis. Even though you can't feel it, you are actually sweating a tiny bit all the time. The sweat comes up through **pores** (pohrs), tiny holes in the skin that allow it to escape. When the sebum meets the sweat, they form a slightly sticky film that both protects the skin and makes objects easier to grasp.

Hypodermis

Underneath the dermis is the deepest layer of skin, called the subcutaneous (suhb-kyoo-tey-nee-uhs) layer, or the **hypodermis** (hahy-puh-dur-mis). The hypodermis consists of a thin layer of fat that cushions your body and helps keep you warm. The hypodermis also attaches the upper skin layers (the dermis and epidermis) to underlying muscles and bones, and supports the nerves and blood vessels located within this layer.

Scanning electron microscope image of fat cells from the hypodermis (300x).

Nails

Fingernails and toenails grow out of the top level of the skin, the epidermis. The hard surface of our nails have important jobs to do: help protect the tips of our fingers and toes, and make it easier to scratch an itch. Nails are made of a protein called keratin (kair-uh-tin), which makes them hard yet pliable. It is the same substance that forms hair, but the keratin in nails is super condensed and does not have any pigments to make color. In addition to offering protection, the nails stiffen the ends of the finger, making them more able to perform delicate actions such as turning the pages of this book.

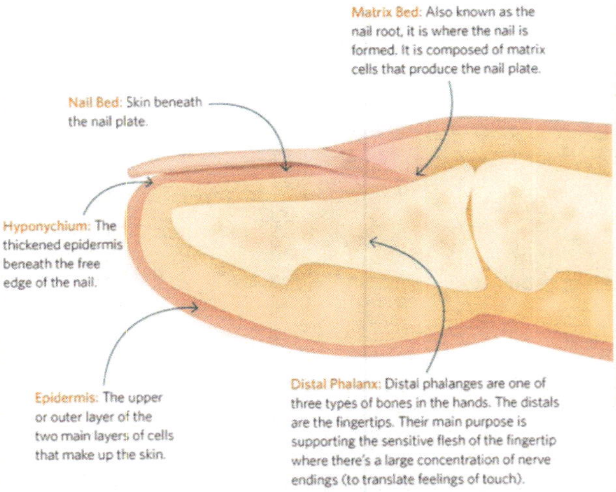

Fingernail anatomy

The place where the nail meets the skin is called the **cuticle** (kyoo-ti-kuhl). The cuticle helps protect the new nail as it grows. It may look like the nail is growing out of the cuticle, but it actually starts in the root of the nail, hidden under the cuticle. When the body forms new cells at the root of the nail, the new nail cells push out the old ones. Thanks to the keratin made by these cells, the old cells flatten and harden. The newly formed nail then inches along the nail bed, the flat surface under your nails, as your fingernails grow. The nail bed sits on top of tiny blood vessels that feed it and make your nails look pink. Really, the true color of the nail is white, but you only see the white when your fingernails grow out past your fingers.

After learning this section, it is hard not to be really impressed at the brilliant designs found in the various components of skin. If we weren't so used to our skin, it would surely be considered a suit with super powers. While the analogy is applicable to just about all of your organs, to appreciate skin's worth, imagine that someone offered you a billion dollars for your skin. Would you take it? If the answer is no, then you can get an idea of just how valuable skin is and what a great gift was given to you by your Creator!

If someone gave you a billion dollars today, would you say thank you?

HEALTHY HABITS:

Skin Care

Skin is a part of the body that is constantly at work. Some people put a lot of time and energy into keeping their skin healthy and nice looking, but the truth is that the most effective way of preventing skin problems and keeping your skin looking great is to make healthy lifestyle choices. Here are five healthy choices that are sure to help keep your skin in tip-top shape.

The first is to protect yourself from the sun. A lifetime of sun exposure can cause wrinkles, age spots, and other skin problems, as well as increase the risk of skin cancer. Ways of protecting yourself from the sun include: using sunscreen, seeking shade during the brightest times of the day (10:00 a.m. to 4:00 p.m.), and wearing clothes that cover as much of your skin as possible—such as long-sleeved shirts.

Another important one is to avoid smoking, since it makes skin look older and contributes to wrinkles. Smoking also narrows the tiny blood vessels in the outermost layers of skin, which decreases blood flow and makes skin paler. This depletes the skin of oxygen and nutrients that are important to skin health.

A well-balanced diet provides the energy and raw materials needed for the growth and replacement of hair, nails, and skin cells. In addition to eating healthy, drinking plenty of water helps replace the water lost in perspiration. Keeping your skin hydrated is also important—that includes using moisturizer when your hands are dry and limiting long, hot baths and showers, since they remove natural moisturizing oils from your skin.

Stress has also been said to make skin more sensitive and trigger acne breakouts and other skin problems. To encourage healthy skin, and a healthy state of mind, take steps to manage your stress. Get enough sleep, set reasonable limits about how hard and long you will work on a project or study for a test, and above all, remember that Hashem loves you and will provide you with everything you need.

SECTION 4.1 REVIEW

1. What is included in the integumentary system?

___

2. What are the five main functions of skin?

___

3. What are the three layers of skin?

___

4. How does the skin help regulate body temperature?

___

5. In what ways do nails benefit you?

___

USE YOUR MIND

Why do you think Hashem made it that skin cells are replaced so often?

___

___

SECTION 4.2: BONES

Have you ever seen a building under construction that had a visible steel frame? Once the building is finished, the steel frame can no longer be seen, but it is that frame that holds the building up. In our body, bones are the framework that holds us together and enables us to stand up straight—which is a unique trait not found in any members of the animal kingdom.

Your bones are like these beams, but they're alive and they do a lot more than just hold you up.

All of the bones together are referred to as your skeleton or **skeletal system**. Although the skeletal system provides shape and support to the body, like the metal beams of a building, it performs other important tasks, such as enabling us to move, protecting our organs, producing blood cells, and storing minerals and other substances like calcium and phosphorus that are released when needed. Because they are solid and hard, it is easy to overlook the fact that bones are complex living structures that grow and develop like other parts of your body.

Bone Structure

About 70% of your bones is not made of living tissue, but rather of hard minerals like calcium. There are three main layers in bones: compact bone, spongy bone, and bone marrow. In addition, on the outside of bones there is a thin, tough membrane called the **periosteum** (per-ee-os-tee-uhm), which covers all but the ends of bones. Blood vessels and nerves that service the bones enter through the periosteum. Underneath the membrane lies a layer of **compact bone** that is hard and dense, but not solid, as a series of small canals runs through it. Nerves and blood vessels run through these canals, supplying the living cells within the bones with what they need.

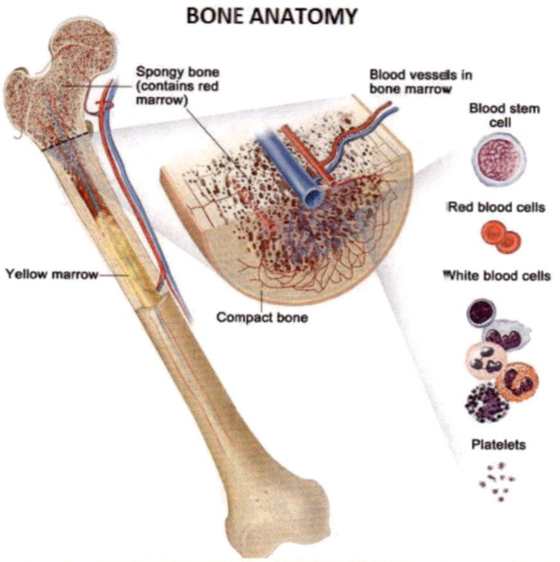

Just under the compact bone is a layer of **spongy bone**, which gets its name due to the presence of many small spaces that make it resemble a sponge. The spongy design enables bones to be lighter, so they don't weigh us down while still maintaining their strength. In addition to this inner layer, the ends of bones are exclusively made up of spongy bone.

Inside many of the larger bones, such as those of the spine, hips, and shoulders, there is also a soft tissue called **marrow**. Marrow is a thick jellylike substance. There are two types of marrow, red and yellow, which both have very important tasks. Red marrow produces most of the body's blood cells, and yellow marrow stores fat that can serve as an energy reserve in times of need.

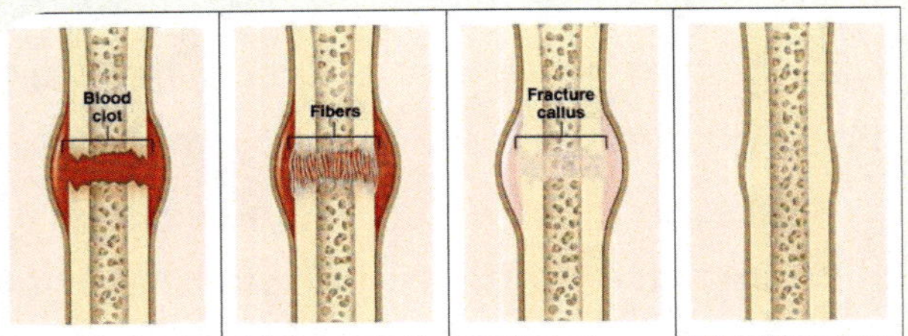

Tissues bleed around the fracture. This forms a blood clot in the space between bone fragments.	Cells form a network of strong fibers inside the blood clot. These fibers hold bone fragments together.	The fibers are replaced by new bone. At first, the new bone is weak and spongy. This is called a fracture callus.	The new bone grows stronger, even after a cast is removed. The fracture callus shrinks as the bone is used.

Stages of bone repair following a fracture

After the age of about 25, bones stop growing, but in children, new bone cells are formed all the time. That causes them to grow bigger and taller. Sometimes, though, new bone forms in adults, for example to replace damaged or broken bone cells. Bone cells can break or become damaged due to an injury, such as a broken or fractured bone, or even when playing sports or by jumping from a high place if the force of the leap causes tiny fractures in the bones. When this happens, the bone-building cells that may have been long inactive are called back into action and get to work building new bone that repairs the fractures.

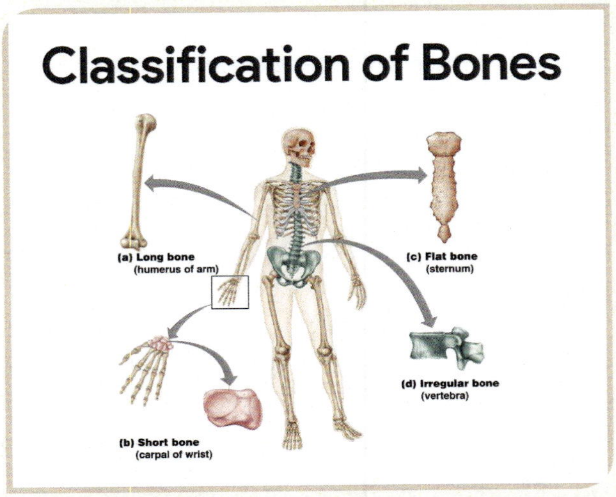

Depending on their shape, bones are classified as long, short, flat, or irregular. Long bones are easy to spot because they are longer than they are wide. Finger bones, arm bones, and leg bones are all good examples of long bones. Short bones tend to look like a cube. The bones in your wrists and ankles are short bones. Flat bones are thin and look flattened. Examples include the sternum (the bone down the middle of your chest that your ribs are connected to), shoulder blades, and the pelvic (hip) bones. Irregular bones have weird shapes and can be found all over the body. The bones in the spinal column are irregularly shaped.

Joints

Since bones cannot bend without breaking, something else is needed to allow the human body to move. To enable movement, Hashem made joints. A **joint** is where two or more bones meet, and it allows movement between those bones. The amount of movement that can occur depends on the type of joint.

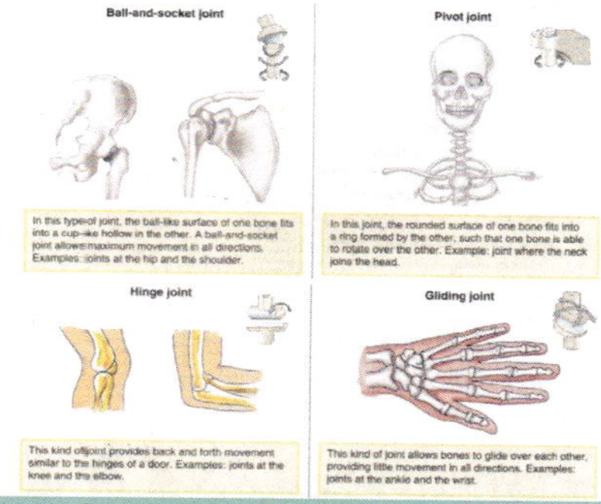

Four of the most common movable joints in the body

Joints allow our bodies to move in many ways. Some joints open and close like a hinge (such as knees and elbows). Others allow for more complicated movement; a shoulder or hip joint, for example, allows for backward, forward, sideways, and rotating movement. Joints are classified by their range of motion. Immovable joints connect bones in a way that allow little or no movement. This is the case in places that you don't want the bones to move around much, such as the skull, which protects our delicate brain.

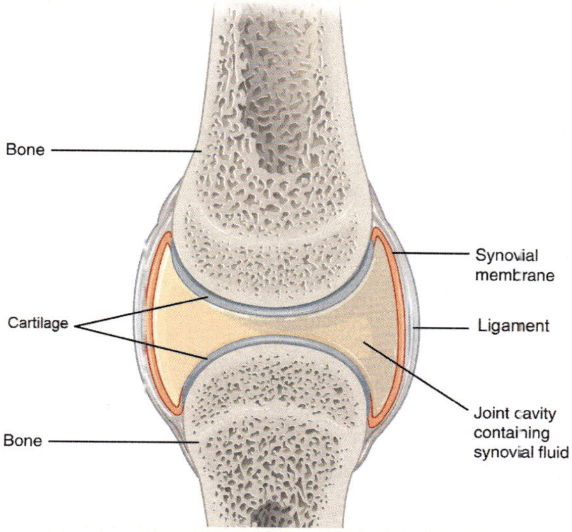

Most of the joints in the body are movable joints that allow the body to make a wide range of motion. The bones in movable joints are held together by strong connective tissues called **ligaments** (lig-uh-muhnts). Ligaments keep the bones firmly attached to one another while still allowing them the flexibility to move to various degrees. To further improve mobility, most joints have a second type of connective tissue called **cartilage** (kahr-tl-ij). Cartilage covers the ends of bones and keeps them from rubbing against each other, since that would be painful. In addition, the ends of bones are lubricated by a thick fluid, called synovial (si-noh-vee-uhl) fluid, which has a similar look and consistency to an egg white and allows bones to move smoothly past each other.

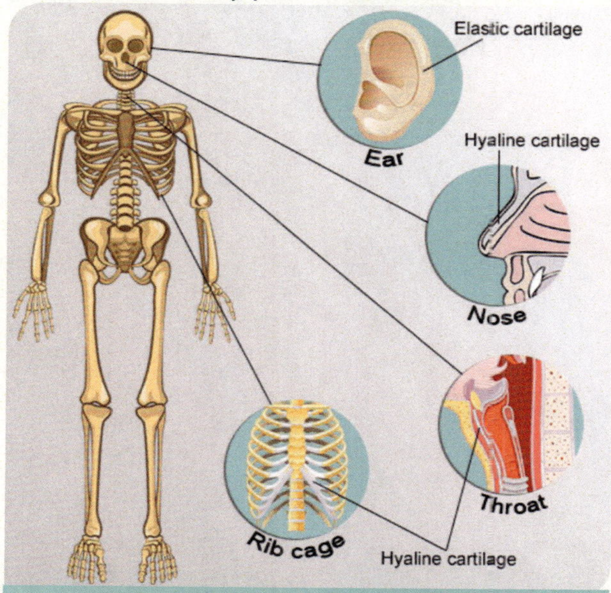

Places in the body that contain cartilage

If you want to feel some cartilage, try wiggling the tip of your nose. It's bendable due to the presence of cartilage. It is interesting to note that all bones start out as cartilage. As children grow, most of the cartilage eventually gets replaced with bone. Thankfully, this process takes time, and therefore some of the elasticity remains throughout childhood. Hashem planned it this way since He knows children have a greater tendency to fall. Young, flexible bones are less likely to break than adult ones.

Spine

Certain major bones in your body are worthy of special attention, such as those that make up the **spine** (spahyn). The spine is one part of the skeleton that's easy to feel. Try reaching around to the center of your back and you'll feel its bumps under your fingers. The spine lets you twist and bend, and it holds your body upright. It also protects the spinal cord—a large bundle of nerves that sends information between the brain and the rest of the body (see chapter 5).

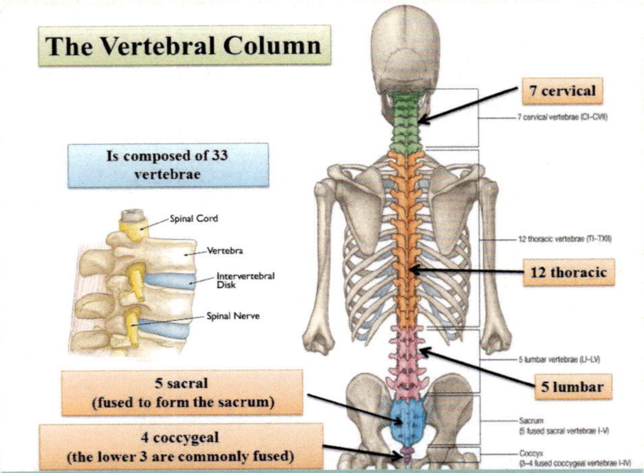

The spine is also referred to as the vertebral column and it has five different types of vertebrae.

The spine is made up of 33 bones called **vertebrae** (ver-tuh-bray; singular: vertebra) and each one is shaped like a ring. In between each vertebra are small disks made of cartilage. These disks keep the vertebrae from rubbing against one another, and they also act as the spine's natural shock absorbers. When you jump in the air or twist around, the disks give your vertebrae the cushioning they need.

Ribs

Our heart, lungs, and liver are all very important, and they are kept safe thanks to the protective rib bones Hashem made to encase them. **Ribs** act like a cage of bones around the chest. It's easy to feel the bottom of this cage by running your fingers along the sides and front of your body, a few inches below your heart. If you breathe in deeply, you can also easily feel your ribs right in the front of your chest.

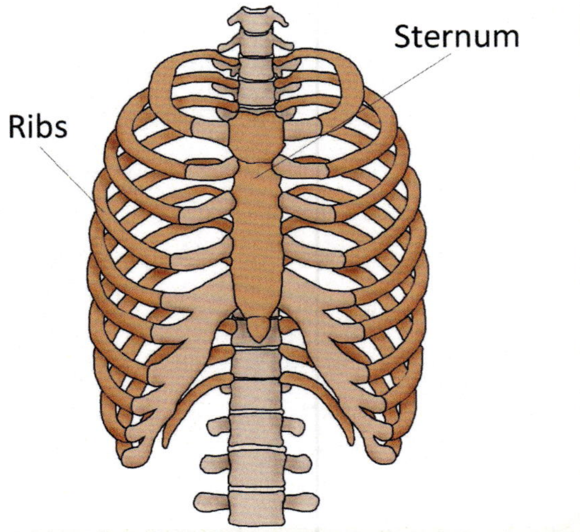

Ribs come in pairs, and the left and right sides of each pair are exactly the same. Most people have 12 pairs of ribs, but some people are born with one or more extra ribs, and others might have one pair less. All 12 pairs of ribs attach in the back to the spine, where they are held in place by vertebrae. In the front, however, only the first 7 pairs of ribs attach to the sternum, a strong bone in the center of your chest that holds those ribs in place. The remaining sets of ribs don't attach to the sternum directly. The next 3 pairs are held on with cartilage to the ribs above them. And the bottom 2 sets are called "floating ribs" because they aren't connected to the sternum or the ribs above them, but instead stay in place simply due to their attachment to the spine in the back.

Skull

The **skull** (skuhl) protects the most important body part of all, the brain. You can feel your skull by pushing on your head, especially in the back a few inches above your neck. The skull is not one continuous bone, but actually made up of different types of bones that are attached to each other. Some of these bones protect your brain, and others make up the structure of our face. The lower jawbone is the only bone in the head that we can move. It opens and closes to let us talk and chew food.

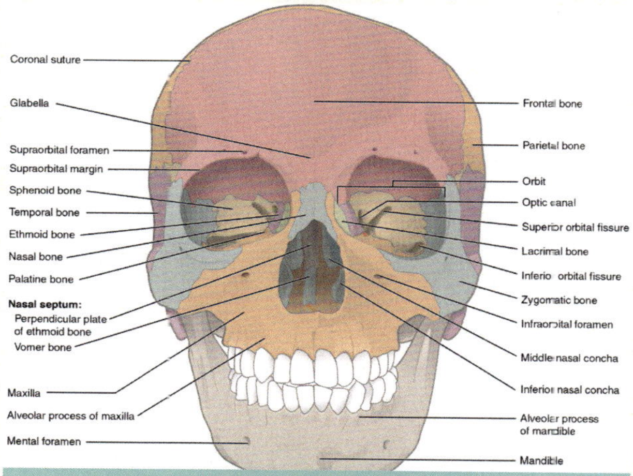

Bones of the skull

Your skull is pretty cool, but it's changed since you were a baby. All babies are born with spaces between the bones in their skulls. As the baby grows, the spaces between the bones slowly close up and disappear, and special joints called sutures (soo-chers) connect the bones.

Hands

When you write with a pen, swing on a swing, or even just pick something up, you are using the bones in your fingers, hand, wrist, and arm. Each arm is attached to a shoulder blade, a large triangular bone on the upper back corner of each side of the rib cage. The arm is made up of 3 bones that are all wider at the ends. That gives them added strength in the spots where they meet another bone.

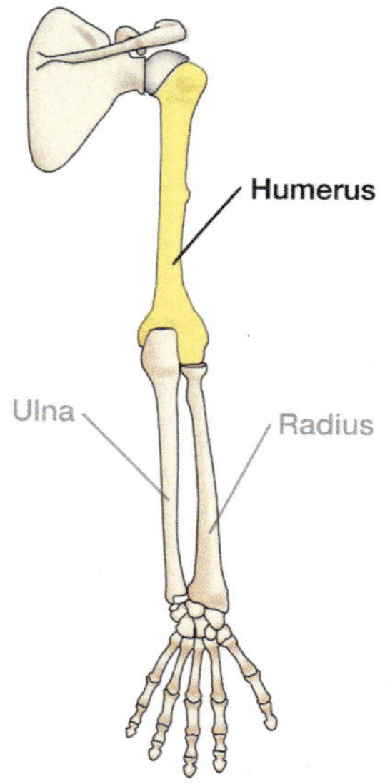

Bones of the arm

At the end of the arm are 8 smaller bones that make up your wrist. Although these bones are small, they have an extraordinary range of motion! Twist your wrist around or wave and you'll see how the wrist can move. After the wrist, the center part of your hand is made up of 5 separate bones. Each finger on your hand has 3 bones, except for your thumb, which has

2. So between your wrists, hands, and all your fingers combined, you have 54 bones that were designed to work together to help you grasp things and carry out all of your needs.

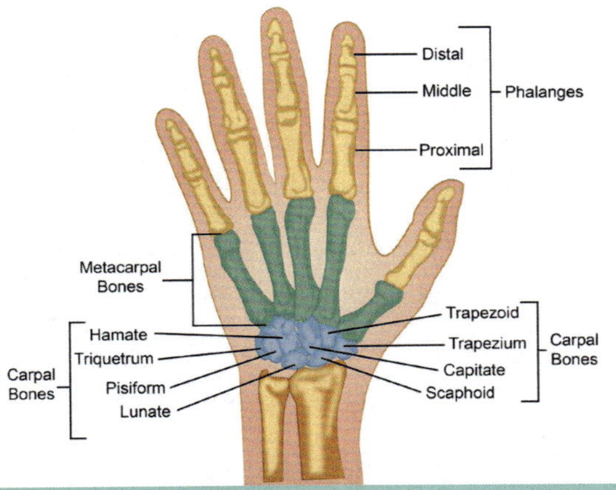

Bones of the hand

Legs

The bones of your legs were designed to help you walk around and run when necessary to carry out your daily activities and do mitzvos. Your legs are attached to a circular group of bones called the **pelvis** (pel-vis). The pelvis is a bowl-shaped structure that supports the spine. In addition to providing support, the pelvis acts as a tough ring of protection around some of the delicate structures in the lower body, such as the bladder, and connects the upper part of the body to your legs.

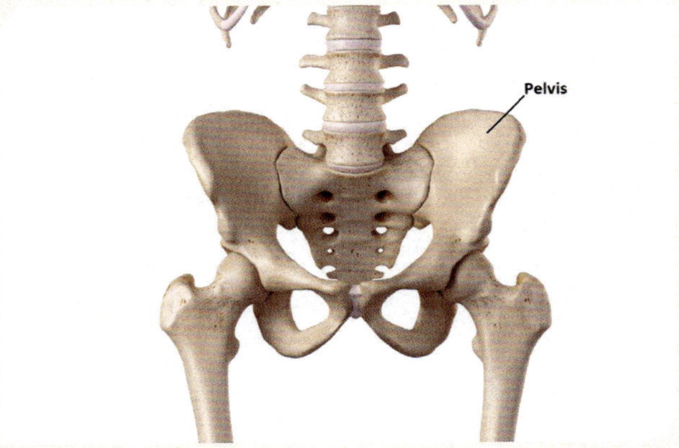

The bones in the legs are very large and strong in order to help support the weight of the body. The bone that goes from the pelvis to the knee is called the **femur** (fee-mur). It's the longest bone in your body. At the knee, there's a triangular-shaped bone called the patella (puh-tel-uh), commonly referred to as the kneecap, which protects the knee joint. Below the knee are 2 other leg bones. Just like the 3 bones in the arm, the 3 bones in the leg are wider at the ends than in the middle to give them extra strength.

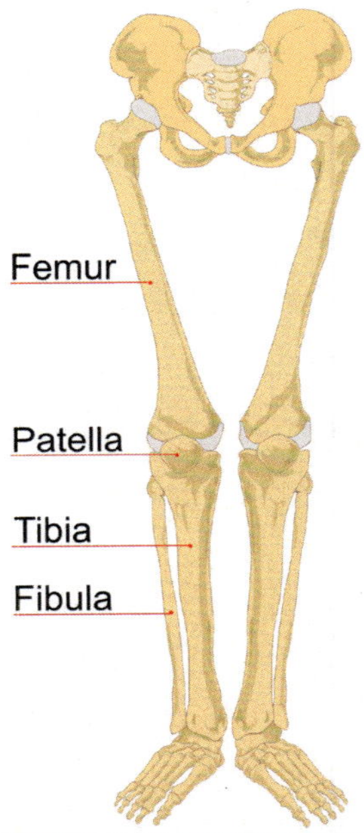

Bones of the leg

While the structure of the ankle, which has 7 bones, is a bit different than that of the wrist, the main part of the foot is indeed similar to the palm of the hand, with 5 bones. Each toe has 3 tiny bones, except for the big toe which has just 2. This brings the total number of bones in both feet and ankles to 52.

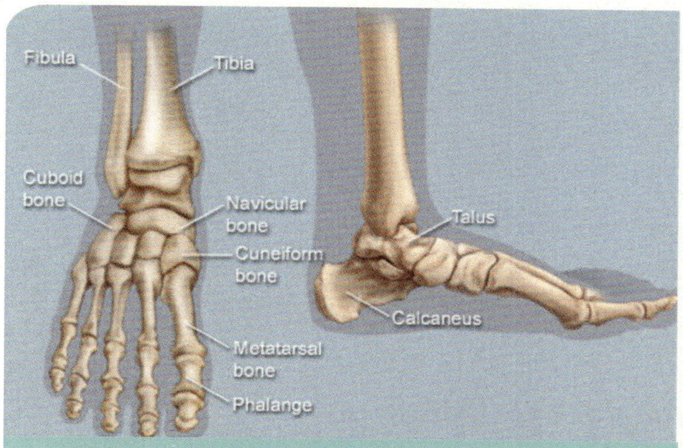

Bones of the feet

The toes and feet are not often used for delicate tasks like picking up objects, but they are frequently used for two very important tasks: standing and walking. If all the bones of the foot did not work together, it would be impossible to balance properly. The bones were arranged perfectly so the foot would be almost flat and a bit wide, to help us stay upright. So the next time you're walking, be sure to look down and thank Hashem for your feet!

HEALTHY HABITS:
Building Better Bones

Since about 70% of our bones are not living tissue, but are made up of hard minerals like calcium and phosphorus, it is really important to eat a well-balanced diet. It is pretty ease to get enough phosphorous with just about any diet today, but many people do not get the amount of calcium that they need for their bones. This is especially true in the teenage years, when you are growing a lot. Meats, whole grains, leafy green vegetables, and dairy products are all rich in calcium.

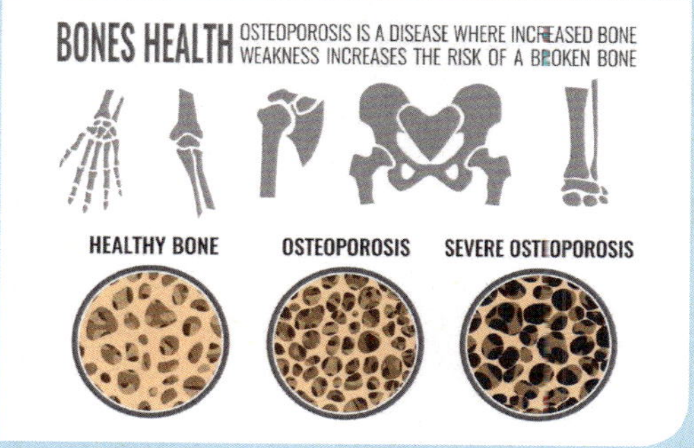

The other major factor in building better bones is to get plenty of exercise. When you run or do other forms of exercise, your bones grow stronger and harder. In addition to building healthy bones, regular exercise and getting sufficient calcium can prevent a condition that some people get when they become older called **osteoporosis** (os-tee-oh-puh-roh-sis). It is due to the loss of minerals that occurs in older people and causes the bones to become weak and brittle (break easily).

How can a person prevent osteoporosis? Well, Rabbi Shimon taught us in *Pirkei Avos* (2:10) that "the straight path a person should cling to is to see the consequences of one's actions." While a diet high in calcium and exercise will certainly help older people's bones a bit, eating healthy food in your younger years and taking proper care of your body can increase the strength and integrity of you bones so they remain strong even with advanced age. So, think of the consequences of what's on your plate or in your cup…and be nice to your bones!

SECTION 4.2 REVIEW

1. What are the major functions of the skeletal system?

____

2. What is the difference between spongy and compact bone?

____

3. What are the two types of bone marrow and what does each of them do?

____

4. What holds two bones together, and what is the site of their attachment called?

____

5. What is the longest bone in the body?

____

USE YOUR MIND

Why is it so important to eat healthy and exercise as a child in order to have healthy bones when you get older?

____

____

SECTION 4.3: MARVELOUS MUSCLES

Try picking up a heavy object (not too heavy!) and notice what you feel. That straining of your muscles is what enables you to carry out all of the movements you make. Of course you would not get anywhere without the bones that provide a framework for your body, but it is the pushing and pulling of the muscles on the bones that actually gets you moving. Muscles turn energy into motion. They're like the engines or motors of your body.

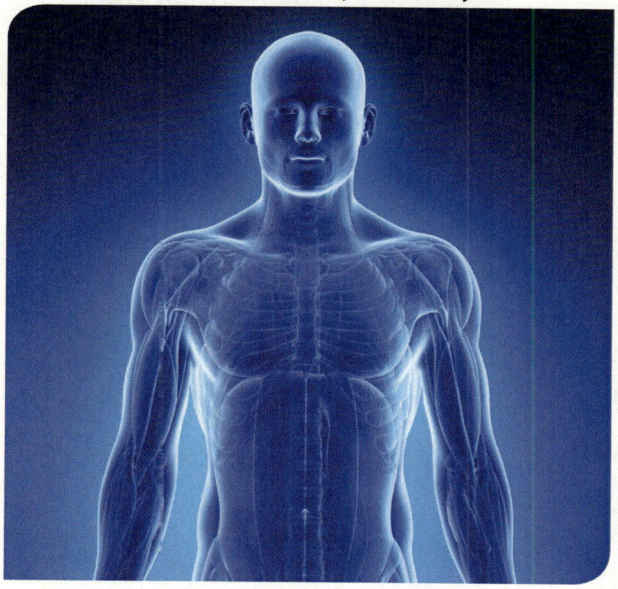

All muscles are made of the same material—a type of elastic tissue that works sort of like a rubber band. Each muscle is made up of thousands, or even tens of thousands, of small, rubbery fibers. The 630 muscles in the body can be divided into three different types: smooth muscle, cardiac (kar-dee-ak) muscle, and skeletal muscle. Together, all of the muscles in the body are referred to as the **muscular system**.

Types of Muscle

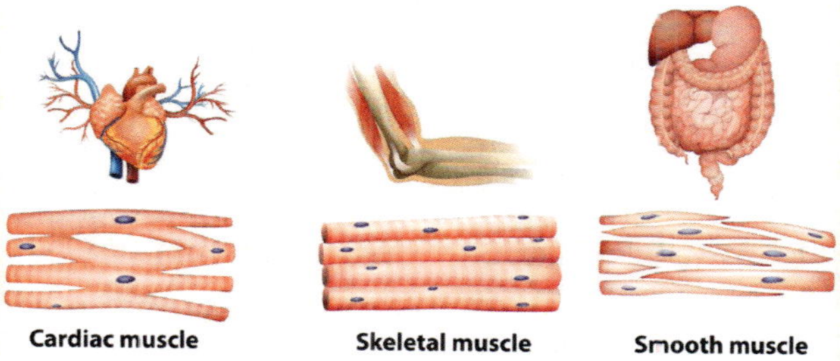

Smooth and Cardiac Muscle

Smooth muscles are sometimes also called involuntary muscles because they contract (tighten up) without you thinking about it. They usually have sheets, or layers, with one layer of muscle behind the other. You can't control this type of muscle. Your brain and body tell these muscles what to do without you even thinking about it. Smooth muscles are at work all over your body, such as in your stomach and digestive system, where they contract and relax to allow food to make its journey through the body.

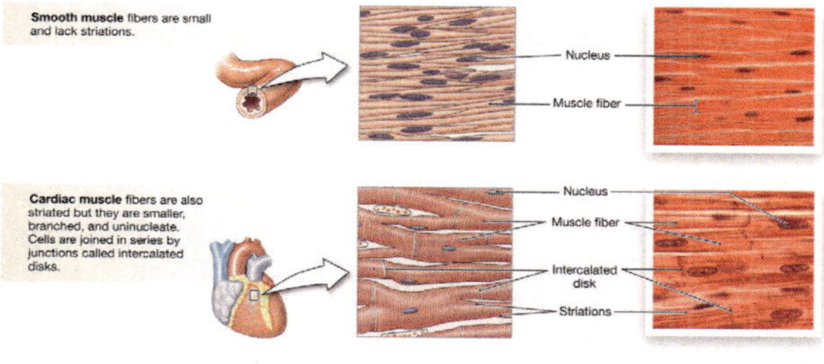

Smooth and cardiac muscle tissues

The muscle that makes up the heart is called **cardiac muscle**. The thick muscles of the heart contract to pump blood out and then relax to let blood back in after it has circulated through the body. Just like smooth muscle, cardiac muscles were designed to work by themselves without any conscious thought. Hashem did this so we could have more time to focus on learning Torah and doing mitzvos, and because He knew their activity is too important to risk us forgetting to tell our heart to pump!

Skeletal Muscle

Now on to the type of muscle we think of when we say muscles. Those are the ones that help us pick up heavy objects or kick soccer balls. These are the **skeletal muscles**, sometimes called **striated** (stry-ay-tud) **muscle** because the light and dark parts of the muscle fibers make them look striped (striated means striped).

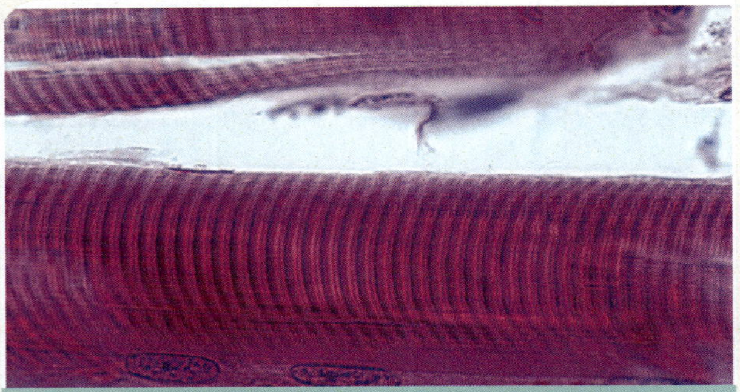

Striated muscle fibers

Skeletal muscles are voluntary muscles, which means you can control what they do. Your leg won't bend to kick a soccer ball unless you want it to. Together, the skeletal muscles work with your bones to give your body the strength it needs to lift objects, walk, and do all of the other physical activities you do. In most cases, a skeletal muscle is attached to one end of a bone. It stretches all the way across a joint and then attaches again to another bone.

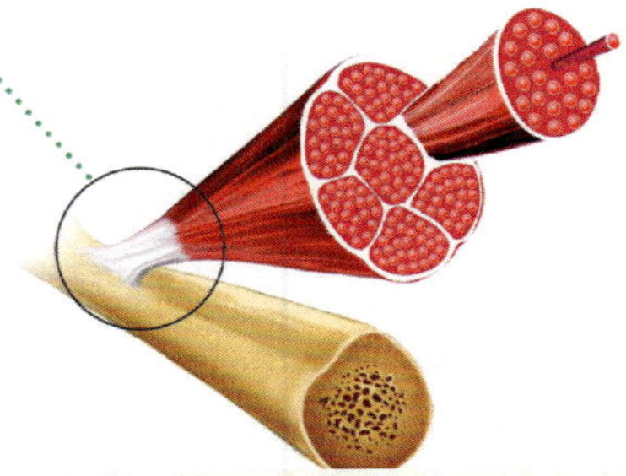

Skeletal muscles are held to the bones with the help of **tendons** (ten-dunz). Tendons are cords made of tough tissue, and they work as special connector pieces between bone and muscle. The tendons are attached so well that when you contract one of your muscles, the tendon and bone move along with it. Skeletal muscles come in many different sizes and shapes to allow you to do many types of wondrous movements, such as turning your head, walking, and jumping up and down. It may seem simple to you, since you do these things every day, but there are literally millions of well-designed "gears" in motion every time you make a move.

How Muscles Work

Muscles work by expanding and contracting. To accomplish this task, they were designed with long, thin cells that are grouped into bundles or fibers. When a muscle fiber gets a signal from a nerve, energy is released by proteins and chemicals. That energy either contracts the muscle or relaxes it. When the muscle contracts, this pulls the bones it's connected to closer together, and when it relaxes, the bones go back to their original positions.

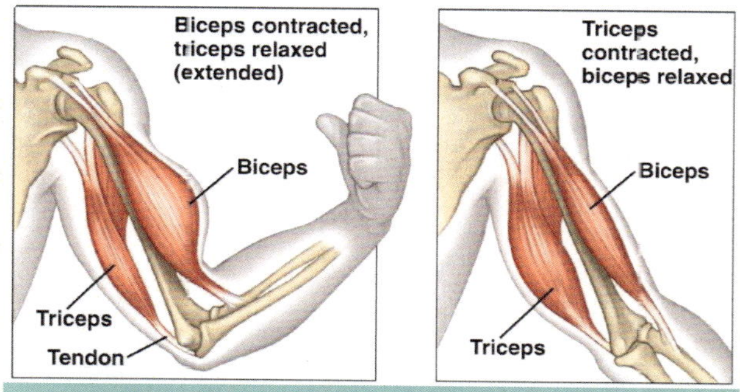

Biceps and triceps contraction

Many of our muscles work in pairs. An example of this is the biceps and triceps in our arms. The biceps have two muscles (*bi* means "two"), and the triceps are made of three muscles (*tri* means "three"). When the biceps contract, the triceps relax, allowing our arm to bend. When we want to straighten our arm back out, the biceps relax and the triceps contract.

Muscle memory helps people master the art of music—if they practice enough.

In addition to the ability to contract when we want them to, Hashem designed skeletal muscles with the ability to "remember" motions that are performed over and over again. This is referred to as "muscle memory." Muscle memory allows us to become more skilled at certain activities, such as sports and music, because the more we practice the more precisely our muscles perform those motions. In this way, we can become experts at skills that were very difficult the first time we attempted them. This is one of the secrets behind the saying "practice makes perfect."

Facial Muscles

There are many noteworthy muscles in the body, but one set of particular interest are those in the face. You may not think of it as a muscular body part, but your face actually has plenty of muscles. You can check them out next time you look in the mirror.

MUSCLES OF FACIAL EXPRESSION

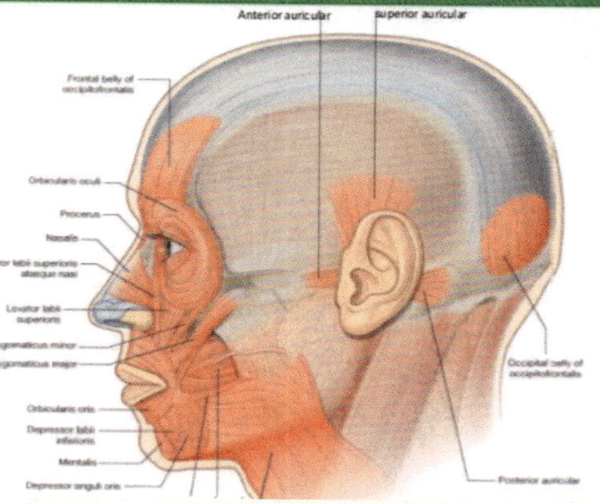

Facial muscles don't all attach directly to bone like muscles do in the rest of the body. Instead, many of them attach under the skin. This allows you to contract your facial muscles even only slightly and make dozens of different kinds of expressions to show how you feel, such as raising your eyebrows to look surprised. To get an idea of how many muscles are working in your face, it takes 17 different muscles to smile and 43 muscles to frown. So, it's a good thing to remember that it's much less work to smile than to frown.

And while you're looking at your face, don't skip your tongue—a marvelous muscle that's attached only at one end! Your tongue is actually made of a group of muscles that work together to allow you to talk and help you break apart food. If you're alone (!) now, stick out your tongue and wiggle it around in front of a mirror to see your marvelous muscles at work.

HEALTHY HABITS:
Making Muscle

Regular exercise is important for making your muscles strong, grow, and be flexible—all of which are important to your health and mobility. When you exercise and put normal strain on your muscles, they actually get bigger. While you certainly need to be careful and not overdo it, if you feel your muscles are sore after exercising, it is a sign that you got a good workout and your muscles are going to grow.

Jews making sure their muscles are at peak performance to do mitzvos

Building your muscles can greatly reduce the natural muscle loss that comes with age, increase bone density (make bones thicker), and reduce the risk of getting injured. Stretching regularly (and especially before exercise) can also bring great benefits, since the more flexible you are, the less likely it is that you will have muscle injuries, such as strained or pulled muscles, which can occur when they are overworked or overstretched. Keeping your muscles strong and flexible can also reduce the risk of torn tendons or "cramped" muscles, which happens when the entire muscle contracts strongly and stays contracted. Muscle injuries can be quite painful, so it's well worth the effort to exercise regularly and enjoy all the benefits healthy muscles can bring.

SECTION 4.3 REVIEW

1. What are the three types of muscles in the body?

___

2. What is the difference between voluntary and involuntary muscles?

___

3. Which types of muscles are voluntary and which are involuntary?

___

4. What are tendons?

___

5. Explain how muscles enable you to move.

___

USE YOUR MIND

Why do you think Hashem made it that people's muscles get stronger the more they use them?

___

___

SECTION 4.4: EXPAND YOUR MIND

SEEING THE PLAN AND THE PURPOSE

When we look at and feel the skin, muscles, and bones, it's perhaps the greatest opportunity for us to see Hashem right in front of our eyes, as the verse says, "From my flesh I see G-d" (*Iyov* 19:26). The importance of taking the time to think about the wonders we see every day was taught by one of the leaders of the previous generation, Rabbi Elazar Menachem Mann Shach, *zt"l*. In one of Rabbi Shach's many discourses on *emunah*, after detailing many of the wonders and divine providence found in nature, he added, "I think about these things all the time."

If you stretch out your hand and wiggle your fingers, you are witnessing an incredible, miraculous event that should not be appreciated less simply because it happens so often to so many people. In fact, just the opposite! We should be more impressed that Hashem makes so many miracles every moment. The importance of seeing the daily events of our lives as miracles is explained in the *Sfas Emes* on *Parshas Behar-Bechukosai*, which states, "In truth, *Bnei Yisrael* must know that miracles and nature are one and the same, and in truth there is no miracle greater and more wondrous than that of nature, which is the greatest of all wonders that we can comprehend."

So, when we look at our hands, we might want to think about how they can bend and turn in so many ways, how they are capable of doing the most delicate tasks, such as sewing—and yet when needed for self-defense can produce a powerful brute force capable of causing serious damage. We can think about the millions of sensory receptors that are tell us if objects are hot or cold, hard or soft. And of the fact that the skin is part of a carefully regulated temperature-control system that can quickly flood the extremities with blood, produce sweat, and cool down the body in the heat.

Walking upright is something we all do, but it's not a simple feat of our feet. There is a delicate balancing act going on; our bones, muscles, and brain are all cooperating with every step we take. While I certainly don't wish it upon any of you, I recently broke my pinky finger and the ordeal was quite unpleasant. But within a short amount of time, my nail grew back, the bone was rebuilt, and now I have a fully functional finger again. Hashem did such a good job of fixing my finger that I cannot even notice there was ever an injury. That is an amazing miracle—a testimony to the wisdom and kindness of Hashem working right before my eyes.

There are many more examples of how to see Hashem in our flesh, so take some time to look around at your skin, think about the benefits you get from your bones, and marvel at how your muscles help you move. Each glance is an opportunity to see an ingenious Divine Design at work.

SECTION 4.4: EXPLORE CHAZAL

TORAS EMES

One of the most famous and mysterious bones, referred to in numerous Torah sources, is the luz bone. The Hebrew word luz literally means "nut" or "almond," and it is perhaps the most important bone in the body—for a surprising reason…

Rabbi Moshe ben Maimon, known as the Rambam, teaches that the last of the Thirteen Principles of Faith is that the righteous who have died will ultimately be resurrected—techiyas hameisim. We know that after death the body decomposes in the ground. If so, how will the dead be resurrected?

The emperor of Rome had this same question and he asked it to Rabbi Yehoshua ben Chanania, whose response is given in Bereishis Rabbah (28:3):

Hadrian [the emperor of Rome] asked Rabbi Yehoshua ben Chanania: 'From where does the Holy One blessed be He reconstruct man in the future to come?' He answered him: 'From the luz of the spine.' Said he: 'From where do you know this?' He said: 'It has come to me [as a tradition], and I will show you.' They tried to grind a luz bone in a mill but it was not ground. They burned it in fire but it was not consumed. They put it in water but it was not softened. They put it on an anvil and he started to beat it with a hammer, but the anvil slipped. The hammer was broken and it remained intact.

There is a difference of opinion as to exactly where the luz bone is located, but all agree it is very small. Rashi says that it' is located at the end of the spine—seemingly a reference to the coccyx (kok-siks), which we call the tailbone. Rabbi Yitzchak Luria Ashkenazi, also referred to as the Ari or Arizal, one of the foremost Kabbalists of all time, taught that the luz bone is located in the place where the knot of the tefillin is placed, meaning where the skull meets the back of the neck. This is generally considered to be the first cervical vertebra, otherwise known as the atlas bone.

In relation to the luz bone, it is interesting to note Dr. Hans Spemann's groundbreaking research on embryology (study of a developing fetus/baby), for which he received the Nobel Prize in 1935. Spemann discovered that the development of humans (as well as other organisms with backbones) begins from a small bone located at the bottom of the spine, the coccyx bone, which he called the "primary organizer," since all the unique aspects of the individual originate in that place.

In a series of experiments conducted by the German doctor and a group of his colleagues, coccyx bones were subjected to stress factors to devitalize (kill) the cells. They were boiled, frozen, dried out, and submersed in alcohol. I doubt they were attempting to prove Rabbi Yehoshua ben Chanania right, but they did discover that such bones were still capable of being transplanted and regenerating new organisms even after being devitalized.[1]

While we certainly do not need scientific verification to support the teachings of our sages, the scientific evidence showing how the luz bone can regenerate organisms "back from the dead" can certainly help to strengthen our emunas chachamim and belief in techiyas hameisim.

I believe with complete faith that there will be a resurrection of the dead at the moment such a desire emanates from the Creator. May His Name be blessed and His mention be exalted, forever and for all eternity.

1 Navis, Adam R., "'Versuche zur Analyse der Induktionsmittel in der Embryonalentwicklung' (1932), by Hermann Bautzmann, Johannes Holtfreter, Otto Mangold, and Hans Spemann," Embryo Project Encyclopedia, Nov. 13, 2007.

EXPERIMENT:

DUSTING FOR FINGERPRINTS

Corresponds to Section 4.1

Materials:

1. Hand Lotion
2. Baby Powder (you can also use cornstarch or cocoa powder)
3. Scotch Tape
4. Small Paint Brush
5. Black Construction Paper

Procedure:

1. Rub a small amount of hand lotion onto your hands (this will make your fingerprints easier to find).

2. Press your fingers ont o the table or desk in front of you, leaving behind your prints.

3. Sprinkle a small amount of powder onto the surface you touched.

4. Using the paint brush, gently brush away the excess powder, being careful to leave the fingerprints intact.

5. Place a piece of scotch tape on top of the powdery print, then lift the tape (the fingerprints should have stuck to the tape), and place it onto a piece of black construction paper to see it more clearly.

6. Pick a different place in the classroom and repeat steps 3-5 to see if you can find other people's fingerprints.

Explanation:

Our skin produces natural oils, which leave a residue when our fingers touch or press against a surface or object. The powder sticks to these oils, making visible the unique patterns made by the ridges on our fingertips. No two people have the same fingerprint patterns.

Questions

1. Were you able to find other people fingerprints?

____

2. How do you think fingerprints are useful for detectives solving mysteries?

____

3. In what other ways are fingerprints used?

____

EXPERIMENT:

BUILDING BACKBONES AND FORMING FINGERS[2]

Corresponds to Section 4.2

BUILDING BACKBONES

Introduction

What gives you the ability to bend, twist, run, or skip? Does having more bones in a specific area of the body make us more flexible there? Try this experiment and find out!

Materials

- pipe cleaner
- drinking straw
- scissors

Procedure

1. Thread the pipe cleaner through the straw.

2. Gently try to bend the pipe cleaner where it is covered in the straw. Does the pipe cleaner bend much?

3. Take the pipe cleaner out of the straw and cut the straw into pieces that are about 1 inch (2.5 centimeters) long. Thread the pieces of the straw onto the pipe cleaner so that they are touching each other.

4. Gently bend the pipe cleaner again. How easily does it bend now?

2 From *Home Science Tools*.

Explanation

The pipe cleaner and straw demonstrate how joints allow our bodies to move. When the straw was in just one long piece, it represented one long bone, such as our thigh bone or upper arm bone. These bones can't bend because there is no joint there to allow that to happen. Instead, these solid bones give our bodies stability. But after the straw was cut into pieces and then placed on the pipe cleaner, it was very easy to bend because of all the "joints" created where the pieces of straw meet. Those pieces of straw represent many short bones; a joint is where two or more of them meet.

The small pieces of straw stacked on top of each other are very similar to how your backbone is structured. The spine is made up of small bones stacked on top of each other, with the spinal cord threaded through them. Like the pipe cleaner, you can bend your back forward and backward, side to side, and even rotate. The stacked bone, represented by the pieces of straw, are not very stable, though, so your back has strong muscles to help keep your spine straight.

Your body has a lot of other joints too—bend your arms and legs, wiggle your fingers and toes, sit down, reach up high, and move your head from side to side. It is possible for you to move your body in all of these ways because of joints in your fingers, ankles, knees, hips, elbows, neck, and everywhere else that bones connect inside of your body!

FORMING FINGERS

Introduction

Make a working model of your hand, using paper and string. You'll see how the muscles in your hands pull on your bones to help you move your fingers!

Materials

- card stock or thin cardboard
- pen or pencil
- scissors
- string
- stapler
- tape

Procedure

1. Outline of your hand with the pen or pencil on the card stock or cardboard.

2. Cut out the shape of your hand with scissors.

3. Cut the string into 5 pieces that are each about the length of your hand (from the tip of your middle finger to your wrist.

4. Tie or tape the piece of string to the tip of each finger and the thumb of the card stock and stretch the string it to the base of the palm. Staple the string to the card stock at the same points where you have joints in your fingers and thumb.

5. Try pulling the strings from the base of the palm. What happens?

Explanation

The muscles in your body are there to contract (shorten), a simple but very important task. The muscles inside your forearm (the part of the arm between the elbow and the wrist) have long tendons running toward the wrist. These muscles allow you to bend your fingers toward your palm, such as when you give a thumbs-up. This is what's happening when you pull on the strings of your model hand.

Questions

Pull on different sections of the five strings and see how the cardboard hand moves. Try moving your fingers in the same way. Can you move your own fingers more than the cardboard hand can? Why?

____

____

____

EXPERIMENT:

BUILDING BACKBONES

Corresponds to Section 4.3

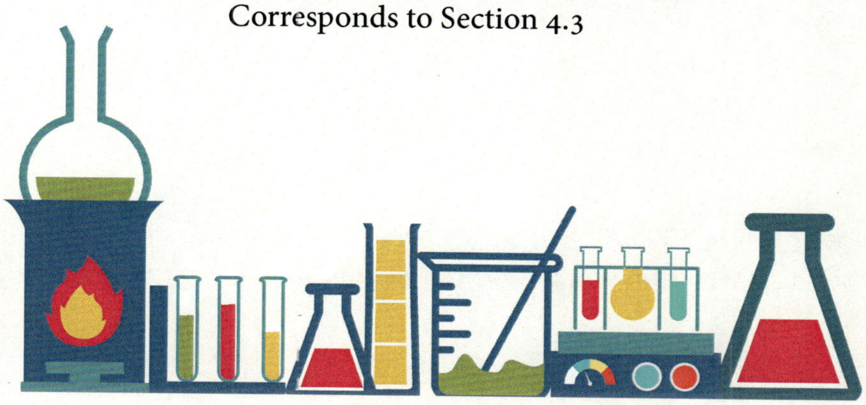

Introduction

Make a working model of your hand, using paper and string. You'll see how the muscles in your hands pull on your bones to help you move your fingers!

Materials

- card stock or thin cardboard
- pen or pencil
- scissors
- string
- stapler
- tape

Procedure

1. Outline of your hand with the pen or pencil on the card stock or cardboard.

2. Cut out the shape of your hand with scissors.

3. Cut the string into 5 pieces that are each about the length of your hand (from the tip of your middle finger to your wrist.

4. Tie or tape the piece of string to the tip of each finger and the thumb of the card stock and stretch the string it to the base of the palm. Staple the string to the card stock at the same points where you have joints in your fingers and thumb.

5. Try pulling the strings from the base of the palm. What happens?

Explanation

The muscles in your body are there to contract (shorten), a simple but very important task. The muscles inside your forearm (the part of the arm between the elbow and the wrist) have long tendons running toward the wrist. These muscles allow you to bend your fingers toward your palm, such as when you give a thumbs-up. This is what's happening when you pull on the strings of your model hand.

Questions

Pull on different sections of the five strings and see how the cardboard hand moves. Try moving your fingers in the same way. Can you move your own fingers more than the cardboard hand can? Why?

____

____

____

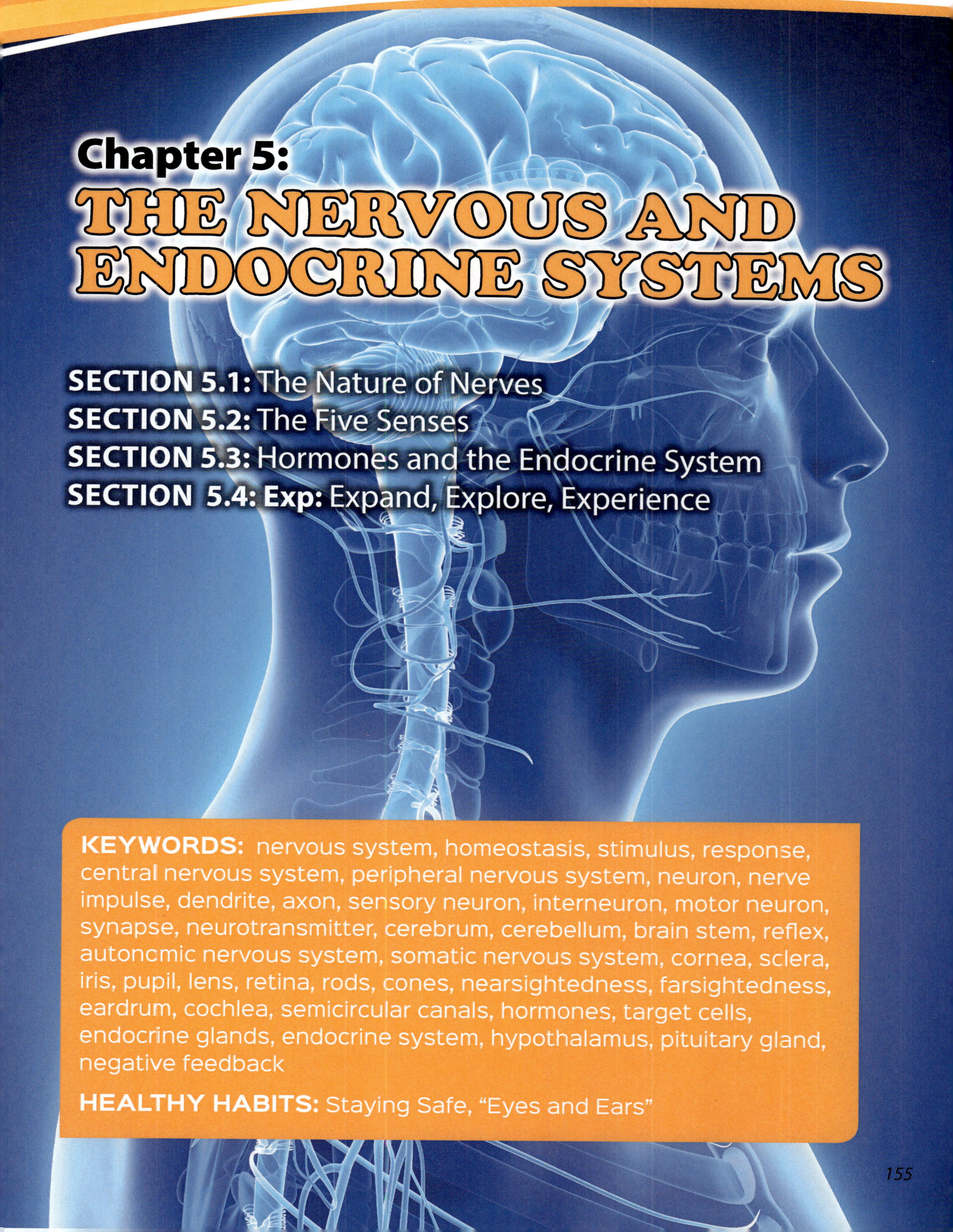

Chapter 5:
THE NERVOUS AND ENDOCRINE SYSTEMS

KEYWORDS: nervous system, homeostasis, stimulus, response, central nervous system, peripheral nervous system, neuron, nerve impulse, dendrite, axon, sensory neuron, interneuron, motor neuron, synapse, neurotransmitter, cerebrum, cerebellum, brain stem, reflex, autoncmic nervous system, somatic nervous system, cornea, sclera, iris, pupil, lens, retina, rods, cones, nearsightedness, farsightedness, eardrum, cochlea, semicircular canals, hormones, target cells, endocrine glands, endocrine system, hypothalamus, pituitary gland, negative feedback

HEALTHY HABITS: Staying Safe, "Eyes and Ears"

SECTION 5.1: THE NATURE OF NERVES

Let's see if you can guess what is really happening in this scenario. It's taking place outdoors, in a field. Someone is inside of a sophisticated machine, shooting a video recording of the events that are occurring there at the rate of about 50 frames per second. The machine has numerous devices that can sense and respond to various forms of input.

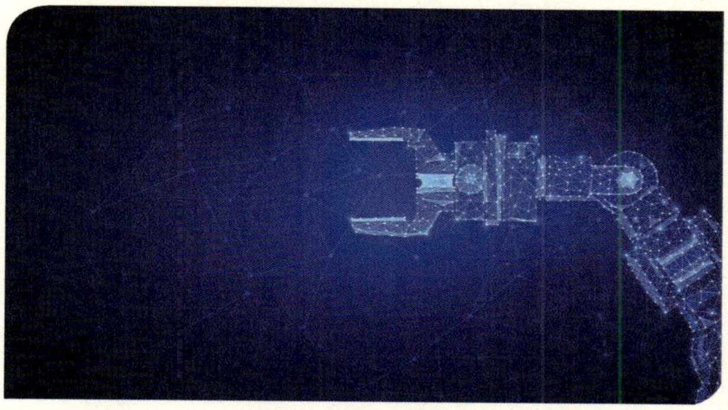

One part of the sensory equipment is triggered to protect the delicate machinery from damage. It responds by releasing an automatic arm, which intercepts a projectile and uses its gripping device to retrieve it. Once retrieved, it signals an advanced targeting system to find the place from where the projectile came and fire it back at the target, launching it high into the air, and causing it to land perfectly, some 53.5 feet (16 meters) away. The trajectory is even more impressive since it had to instantaneously take into account both the distance of the target and a 5-mile-per-hour (8-kilometer-per-hour) wind blowing in the opposite direction.

What we have described above is only a small portion of the numerous tasks performed by your body in order to play a game of catch with a friend in the park. While it is your eyes that do the video recording and your arms that do the catching and throwing, it's actually your nervous system that is processing all of the information coming in. It is directing the body to carry out a game that seems pretty simple—but only because we are unaware of the millions of cellular activities necessary to carry out the various tasks of the game.

Israel's baseball team did surprisingly well in the recent World Baseball Classic, but Hashem deserves the most applause—for designing humans with the ability to catch, throw, run, slide, and keep their eye on the ball!

Functions of the Nervous System

In addition to supplying the sensory equipment the body needs to gather information about what's happening around it, the **nervous system** also includes a network through which your brain sends and receives the information it takes in. Thus the nervous system is able to direct the way your body responds to the information coming in, and help maintain **homeostasis** (hoh-mee-uh-stey-sis), the state of stable internal conditions.

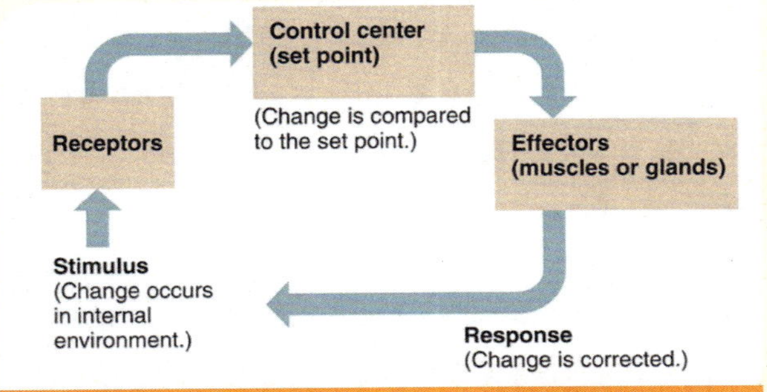

An overview of how homeostasis is achieved in the body

Any activity in the outside environment or within the body that can cause a reaction is called a **stimulus** (stim-yuh-luhs: plural: stimuli). The way the body reacts to a stimulus is called a **response**. Some responses are for the sake of protection, such as moving away quickly from the stimuli of hearing and seeing a buzzing bee. Other responses are in order to maintain homeostasis—such as the nervous system sending the signals of thirst or hunger when your body is low on fuel. The response, eating and drinking, serves to maintain homeostasis by providing the body with the nutrients and energy it needs.

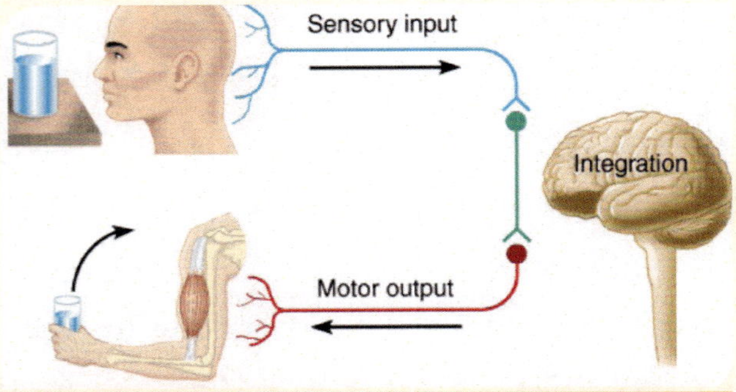

Example of how the brain receives a stimulus and carries out a response to it

Divisions of the Nervous System

Your nervous system consists of two separate divisions, which work together to carry out all of the information processing and directing of bodily processes necessary to survive. The first of the two divisions is called the **central nervous system** and consists of the brain and spinal cord, which act as the control center of the body. All of the information about what is happening in the world outside and inside of you is brought to the central nervous system to be processed.

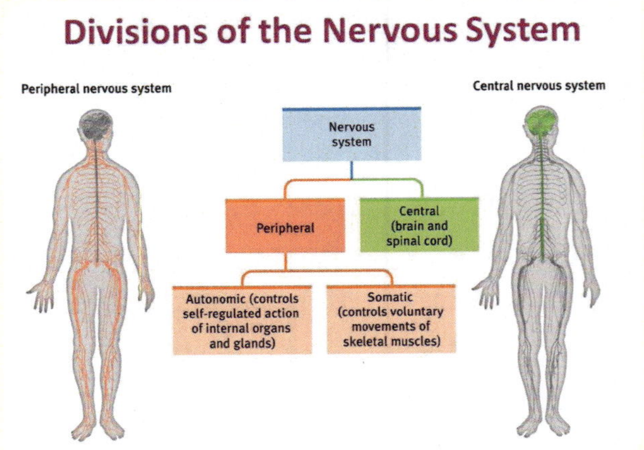

The second division is called the **peripheral** (puh-rif-er-uhl) **nervous system**. It includes all the nerves that branch out from the central nervous system and connect it to the rest of the body. This system consists of 43 pairs of nerves, 31 of which begin in the spinal cord. The remaining 12 originate in the brain. Each pair has one nerve that heads to the right side of the body and one that heads to the left, in order to receive stimuli and send out responses from the central nervous system on both sides of the body.

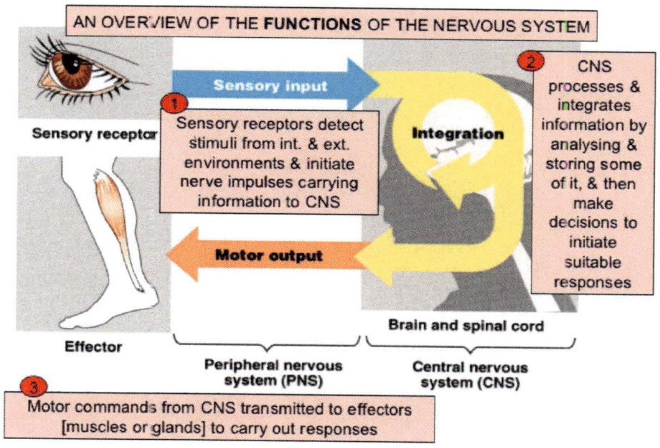

The two nervous systems work together, since the nerves in the peripheral nervous system collect sensory information and send it to the central nervous system for processing. Once a decision is made by the command center as to how to respond, it sends messages back via the nerves to instruct the body to carry out one or more of its possible responses. Some examples of responses include those that are voluntary (happen with conscious thought), such as the legs running quickly when danger is imminent, and those that are involuntary (happen without conscious thought), such as the heart speeding up when more power is needed to carry out strenuous activities.

Neurons

In both the central and peripheral nervous systems, information is gathered and transferred by way of special cells called **neurons** (noor-ons) or nerve cells. The specialized design and structure of the neuron is what enables it to carry messages known as **nerve impulses**.

A neuron has a large cell body that contains the nucleus and other cellular components commonly found in other cells. It also has many short threadlike extensions called **dendrites** (den-drahyts) and a single long extension called an **axon** (ak-son).

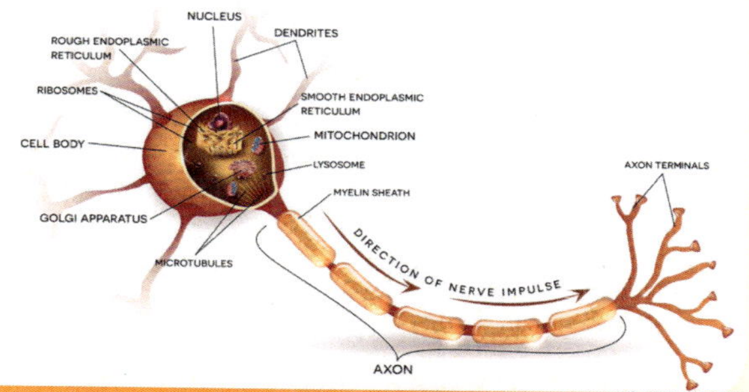

Diagram of a neuron

The dendrites carry nerve impulses toward the neuron's cell body and the axon carries nerve impulses away from the cell—sometimes even far distances away, such as to the sciatic nerve, which starts in the lower spine and goes all the way down to your toes. There are three different types of neurons: sensory neurons, interneurons, and motor neurons.

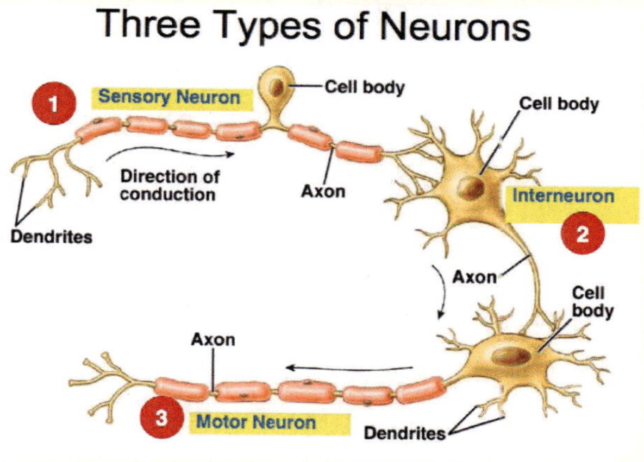

A **sensory neuron** is activated when one of the body's sensory receptors, located at the ends of these cell's dendrites, is triggered to produce a nerve impulse. These nerve impulses are triggered from stimuli such as the pleasure generated from eating ice cream or the pain associated with stubbing a toe.

The dendrites carry the nerve impulse to the cell body. From there, the impulse moves on to the axon—from where it is carried to an **interneuron**. That's a neuron that is designed to carry a nerve impulse from one neuron to another. Most interneurons are located in the brain or spinal cord. It is there that the nerve impulse is processed and a decision is made, either consciously or unconsciously, as to how to react to the stimulus. If a decision is made to react, the interneuron passes the nerve impulse to a **motor neuron**, which sends a nerve impulse to a muscle or gland telling it how to react to the stimulus. An example of such a response might be moving out of the way if danger is sensed.

Path of a Nerve Impulse

Nerve impulses are mostly electrical signals that travel extremely fast. Speeds of nerve impulses range from around 80 miles (129 kilometers) per hour to over 250 miles (402 kilometers) per hour—which is faster than a Formula One race car! While the impulse travels through the dendrites and axons of the nerve cell itself, the impulse is electric. When the impulse moves from one nerve cell to another, however, that occurs thanks to a chemical reaction that happens in a place called the synapse.

A **synapse** (sin-aps) is the junction located between two neurons. It is where the impulse coming from one can be passed on to another. Synapses are located at the end of axon tips. Many axons have multiple tips so they can pass on the nerve impulse to multiple dendrites,

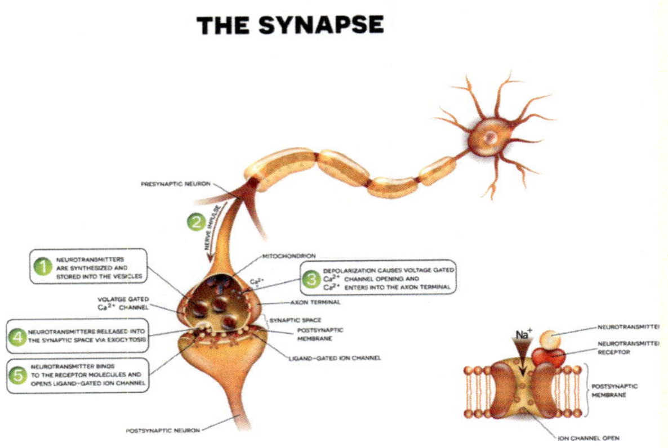

in order to gather more sensory information. While some axons lead to dendrites of other neurons in order to pass along nerve impulses, others lead to muscles or cells in other organs, where they can send signals to perform some type of activity, such as picking up a glass of water when you want to drink it.

When a nerve impulse reaches the synapse, it must cross the space in between the axon and the next neuron that will be carrying on the signal. In order to accomplish this, special chemicals called **neurotransmitters** (noor-oh-trans-mit-ers) are released from storage at the end of axons, into the gap between two neurons. When the neurotransmitters reach the other side of the synapse, many of

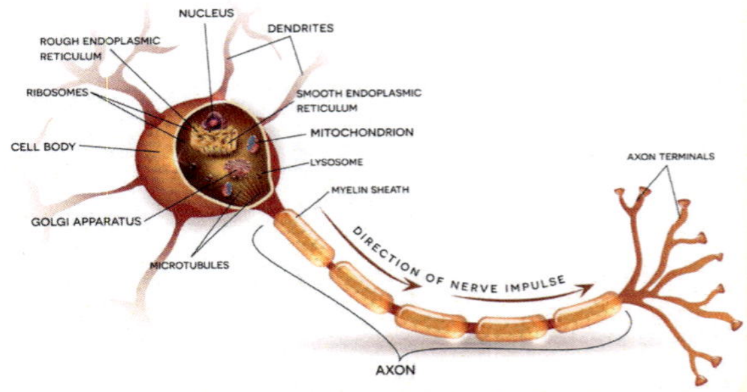

Simulation of neurotransmitters being sent across a synapse

them land on receptors located at the other end (dendrites of the neurons carrying on the signals). The neurotransmitters fit into these receptors like a lock fits into a key, and trigger the next neuron to fire and carry on the signal.

The Brain

If we were to take a vote on what scientists consider the most complex, fascinating, and least understood organ in the body, I think the brain would win. In fact, an article by neuroscientist David Eagleman opened with the statement, "Of all the objects in the universe, the human brain is the most complex."[1]

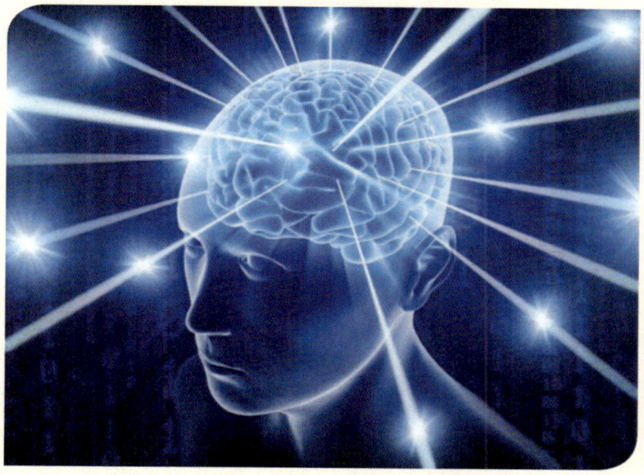

The brain is made up of about 100 billion interneurons that receive, process, and store the information coming to them from the sensory receptors around the body. These neurons are also charged with the task of sending instructions to thousands of other neurons located throughout the body, which are in turn capable of controlling virtually all of the other organs.

Since the brain controls your ability to remember and controls the actions of your body, Hashem made it a hard and sturdy carrying case—the bones in the skull. For further protection and shock resistance, there are also three layers of connective tissue located underneath the skull. These layers are the meninges (mi-nin-jeez) and there is a watery fluid between the middle and innermost layers to aid in shock absorption and protect the brain. The brain itself is made up of three regions that receive and process information: the cerebrum, the cerebellum, and the brain stem.

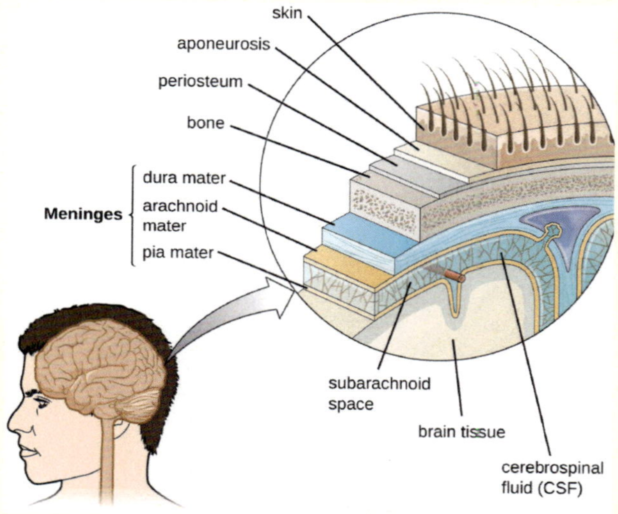

Casings of the brain

Cerebrum, Cerebellum, and Brain Stem

The biggest part of the brain is the **cerebrum** (suh-ree-bruhm). It receives and interprets nerve impulses coming in from the senses, performs complicated mental processes such as learning and remembering, and controls movement. The cerebrum is divided into four distinct regions called lobes. Everything you think about, and the knowledge you acquire as a result, is all in the lobes of the cerebrum. There are distinct areas of the brain that control movements such as speaking or writing, as well as areas that control the senses of touch, taste, smell, hearing, and seeing.

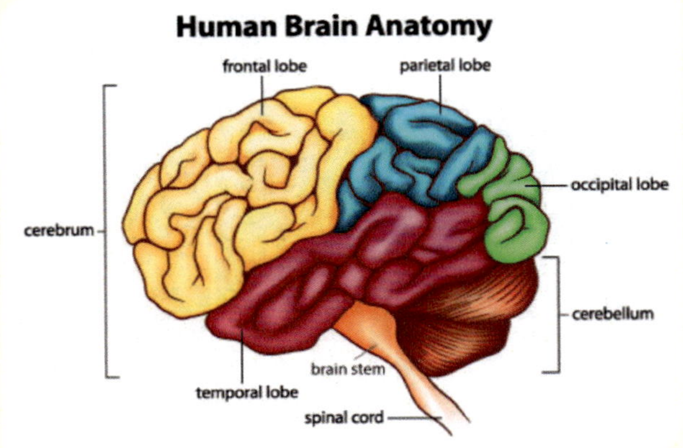

There are two sides to the cerebrum. Each one is responsible for controlling half of the body's movements—but not the one that you'd expect. Ever hear the joke about how only left-handed people are in their right minds? That joke is based on the fact that the right half of the cerebrum controls the left half of the body and the left side controls the right side of the body. In addition, the two halves of the cerebrum carry out different mental activities. The right side of the cerebrum is associated with creativity and art, while the left side is associated with mathematics and logical thinking.

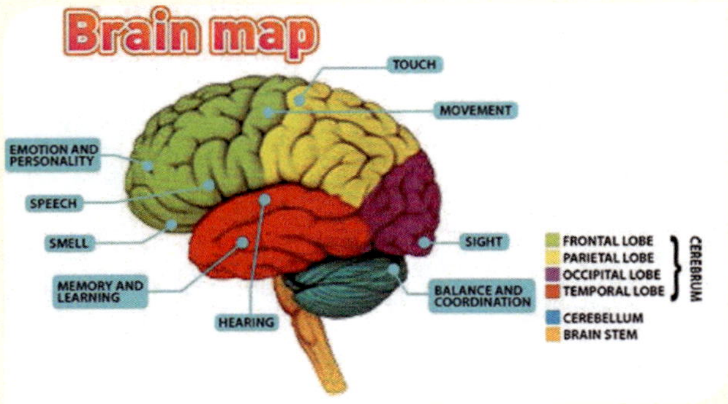

The second biggest part of the brain is called the **cerebellum** (ser-uh-bel-uhm). It is responsible for coordinating the actions of your muscles so you can keep your balance and walk straight. The balancing act of your brain is no simple thing. Humans are the only organism capable of walking upright on two legs. This requires constant vigilance, since everyday activities, such as bending over to pick up a fallen pencil, could easily topple you over—if you didn't have numerous muscles that flex in order to counterbalance the body weight being extended without support.

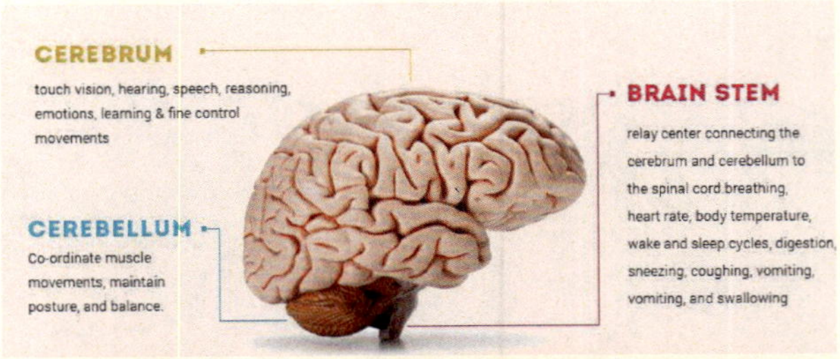

Summary of cerebrum, cerebellum, and brain stem functions

The third section of the brain is the **brain stem**, which controls your body's involuntary actions, such as your breathing and heartbeat, and sneezing. The brain stem is located between your cerebellum and spinal cord. It is also the site that all nerves connected to the brain must pass through in order to carry the signals from the limbs of the body to the brain, and back.

The Spinal Cord

The vertebrae that start in your neck and run down your back, which make up your spine, surround and protect the spinal cord. The spinal cord is the connecting cable. It brings sensory information to the brain and sends back instructions to motor neurons on how to react to what is happening. Like the brain, the spinal cord is also surrounded by three layers of connective tissue and fluid for added protection. The nerves that carry impulses to the rest of the body branch out from the spinal cord and generally affect the organs closest to the vertebrae that they branch out from.

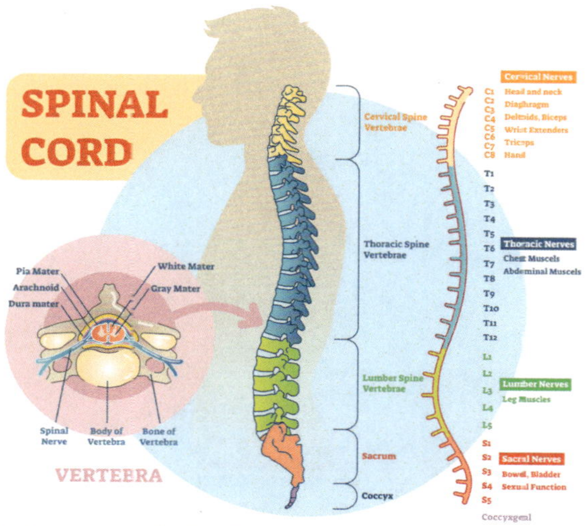

The nerves that service the spinal cord are called spinal nerves. They carry impulses in two directions—from the sensory neurons of the body to the central nervous system; and back to motor neurons, which then cause some sort of movement or other activity in response to what is happening.

Reflexes

While most contractions of skeletal muscles are controlled by the brain, some contract with the involvement of the spinal cord alone. This type of response is referred to as a **reflex**, and the pathway that nerve impulses follow to achieve a reflex is called a reflex arc. Reflexes occur very quickly without any conscious thought. For example, they are frequently triggered when a person touches their hand on a hot or sharp object. At that moment, the sensory neurons send impulses to the spinal cord, from where they are passed to interneurons in the spinal cord itself, not the brain. From the spinal cord, the impulses are passed directly to motor neurons located in the hand, which then recoils from the source of pain.

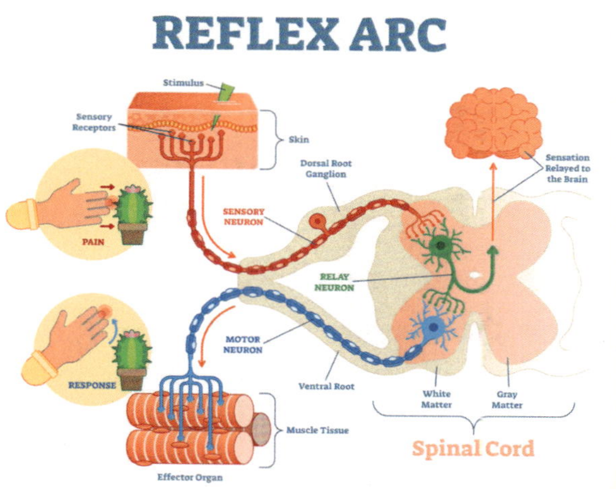

Hashem designed the reflex response for our protection; He wanted people to be able to react as quickly as possible to imminent danger, such as a heavy object that is thrown toward a person, or an intense source of heat that could cause a serious burn. Even though nerve impulses travel very fast, it still takes a tiny amount of time for the impulse to get to the brain and then be sent back out through the motor neurons to respond. By enabling the nerves in the spinal cord to respond directly to impulses announcing extreme danger, our bodies can respond faster, and sometimes that could be the difference between life or death. If you ever do experience a reflex, be sure to do something conscious later—thank Hashem for this built-in security system!

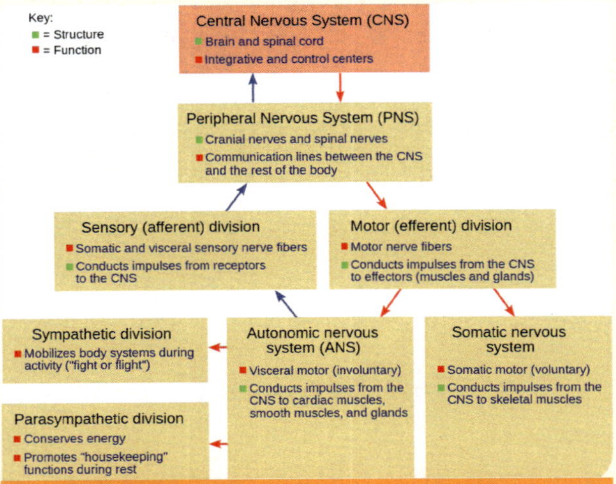

Diagram listing functions of nervous system branches

The nerves located in the peripheral nervous system that cause reflexes are one example of the **autonomic** (aw-tuh-nom-ik) **nervous system** that controls involuntary actions. In contrast, the nerves in the peripheral nervous system that control voluntary actions are called the **somatic** (soh-mat-ik) **nervous system**.

Within the autonomic nervous system, one division of nerves called the sympathetic division creates responses of activity or motion—such as the heartbeat speeding up when danger is approaching and the body prepares to either fight off a foe or flee from them. This is referred to as the "fight-or-flight" response. When the danger has passed or the activity is no longer needed, a separate division of nerves called the parasympathetic division then helps the body restore its equilibrium by relaxing the body and by conserving energy. Both the autonomic and somatic nervous systems work in coordination with the central nervous system to enable you to do all of the incredible actions of your body.

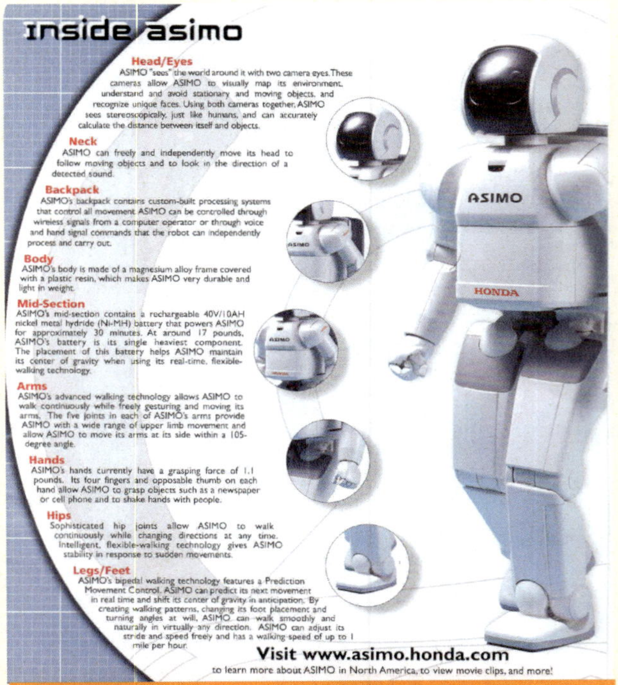

Honda's ASIMO robot is the most advanced humanoid robot today, but is still nowhere near as complex and sophisticated as the miraculous machine you call your body.

It's kind of funny how we are amazed and impressed when we see a robot that can do basic things like picking up a ball and walking around the room, but we overlook the fact that we can do much more complicated tasks. At least once in a while we should stop taking things for granted, and appreciate how we ourselves are live "super robots," more sophisticated than any scientist could ever dream of designing.

HEALTHY HABITS:
Staying Safe

Despite the great number of protective mechanisms that the brain and spinal cord have, a strong enough blow to the head or back can cause serious damage to the central nervous system. Fortunately, there are some basic safety precautions that can prevent most cases of these injuries. The two most common ways the central nervous system can be damaged are concussions and spinal cord injuries. Concussions are injuries to the brain that are very much like bruises you can get on your arms and legs. They occur when the soft tissue of the brain gets pressed against the skull. This can happen when a person hits their head during a hard fall, a car accident, or some other collision, such as those that take place in full-contact sports.

Some concussions are minor and might result in a temporary headache that passes once the bruised brain heals, but others are more serious. With more severe concussions, the accident victim might lose consciousness or feel confused and drowsy afterward.

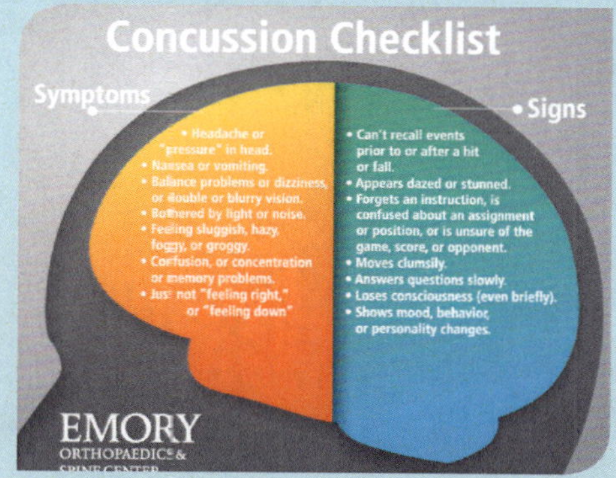

One of the easiest ways to prevent getting a concussion is by wearing a helmet while riding a bicycle, since many instances of concussions come from people riding their bikes without helmets. In fact, the American Association of Neurological Surgeons reports that every year, more than 500,000 people visit emergency rooms in the United States with bicycle-related injuries. In 2009, nearly 85,000 of those were head injuries, and they estimate that up to 85% of head injuries could be prevented through proper usage of helmets.

Spinal cord injuries occur when the spinal cord gets crushed or cut. This results in damaged nerve cells that no longer have the ability to carry impulses through them and pass on information from sensory neurons to the brain. The result is often paralysis—the loss of movement and sensation in a part of the body because the person's limbs aren't getting the messages that they should move.

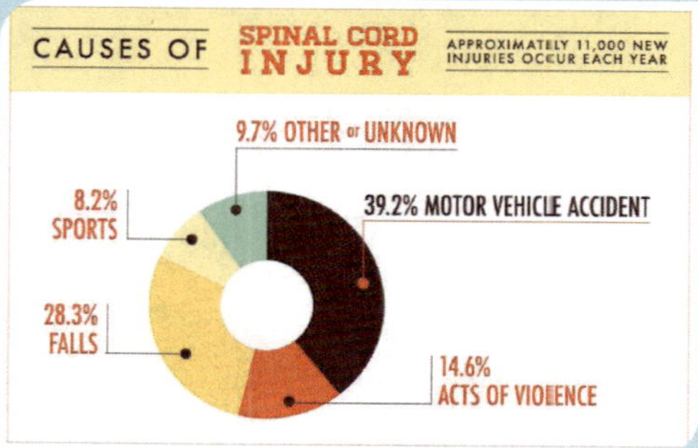

One of the easiest ways to avoid spinal cord injuries is to wear a seat belt while riding in a car. Car crashes are the most common cause of spinal cord injuries, and many of those injuries could have been avoided if the passengers would have been smarter and worn their safety belts. We have already learned that the Torah teaches us, "Be extremely protective of your lives" (*Devarim* 4:15). This can surely be fulfilled by wearing a helmet while riding a bike and a seat belt while riding in a car.

SECTION 5.1 REVIEW

1. What are the components of the central and peripheral nervous systems?

____

2. What are the unique features found in nerve cells that are not found in other cells of the body?

____

3. What is the difference between sensory neurons, interneurons, and motor neurons?

____

4. What is a synapse and how does a nerve impulse travel through it?

____

5. What is a reflex and how is it beneficial?

____

USE YOUR MIND

Injuries to the brain can cause many different symptoms, such as the inability to walk, talk, or remember things. Based on what you have learned, what do you think is the most significant factor that determines the specific effect of a brain injury?

____

____

SECTION 5.2: THE FIVE SENSES

Now that you've learned a bit about the brain, you can start to appreciate how much has to happen for you to sense what is going on in the world around you. All senses come to the brain as electrical impulses carried by neurons. Information about the things we see, hear, taste, touch, and smell all enters the region of the brain that processes it. In what is one of the most complex and least understood processes, that information (in the form of impulses) is miraculously turned into thoughts, feelings, and sensations that enable you to know what is happening in your surrounding environment and interact with it.

Vision

The sense organs that are making it possible for you to read the words in this textbook right now are the eyes. Your eyes were designed to respond to the stimulus of light and send impulses for the brain to interpret so you can see images. It is a true wonder when you think about how the things you "see" are actually only virtual images created by your brain—images that were reconstructed based on the patterns of incoming light. This is similar to how a camera records images on light-sensitive data chips that can be viewed later by programs that read and interpret the information.

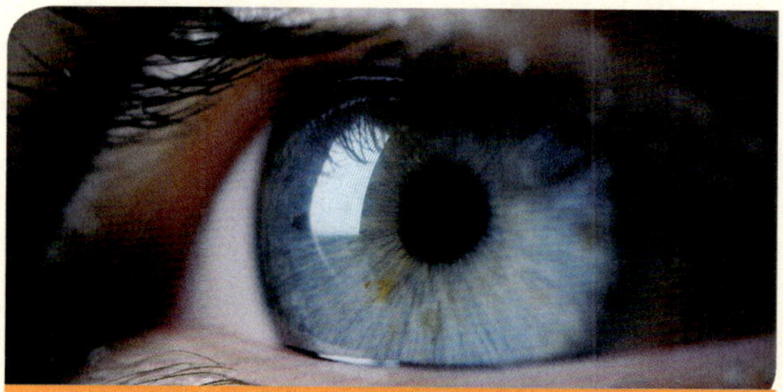

Eyes are our windows to the world.

There are a few key structures that enable the eye to take in light and record images based on their color, pattern, and intensity. The first structure light meets when entering the eye is the **cornea** (kawr-nee-uh). That is a dome-shaped structure made of tissue that is as smooth and clear as glass, yet strong and durable. The cornea was designed to carry out two important functions. The first is to shield the eye from germs, dust, and harmful substances. The second is to project the incoming light properly, so images you see appear clear and recognizable.

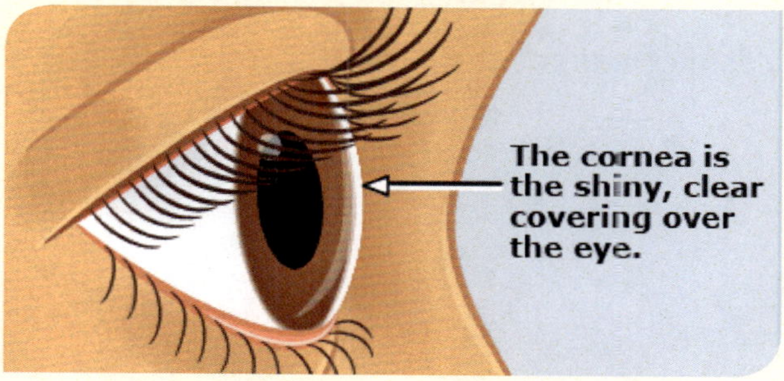

Directly behind the cornea is a chamber filled with a clear fluid called aqueous (ey-kwee-uhs) humor, which supplies the area with nutrients and oxygen. This fluid essentially does the same jobs as blood, but Hashem made it transparent (see-through) because if our eyes were almost all red, that would distort the images—and make everybody look really tired too!

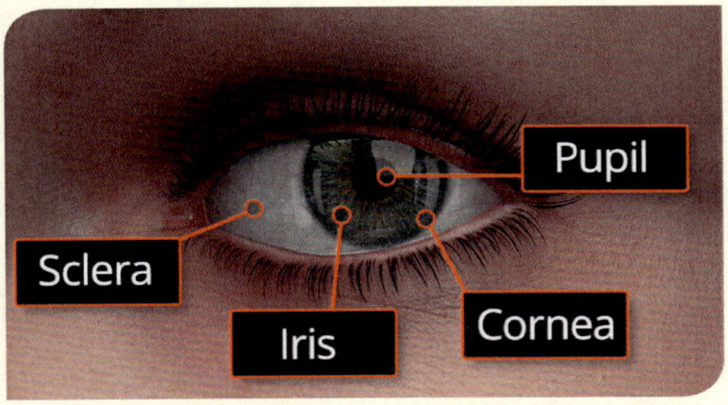

The **sclera** (skleer-uh) is the white part of your eye. It's a tough, protective covering, and the muscles that control eye movement are connected to it. Within the sclera lies the **iris** (ahy-ris), which is the colorful part of the eye. In addition to making your eyes look nice, the iris has muscles attached to it that can change its shape, making it larger or smaller and control how much light gets through the pupil. The **pupil** (pyoo-puhl) is the black circle in the center of your eyes. It dilates (gets larger), when there is not much light—like during the night—and undilates (gets smaller) when there is a lot of light around—like on a sunny day or when someone points a flashlight at you.

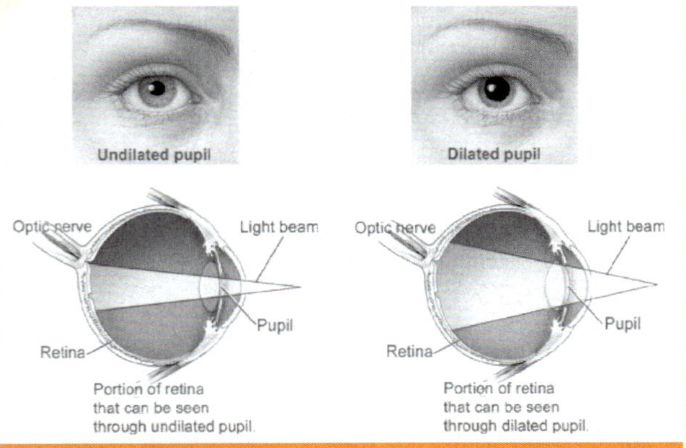

Dilated vs. undilated pupils

The pupil gets bigger when less light is available in order to be able to take in as much of the existing light as possible, so that you can still see. You may have noticed that sometimes when you first enter a dark room you can't see anything at all, but after a while you can see a little. That's because the pupil is busy opening wider so you can manage in the dark. Then, when it's bright, you don't need a lot of light—it could even be potentially harmful—so the pupil shrinks to keep out light that is not necessary.

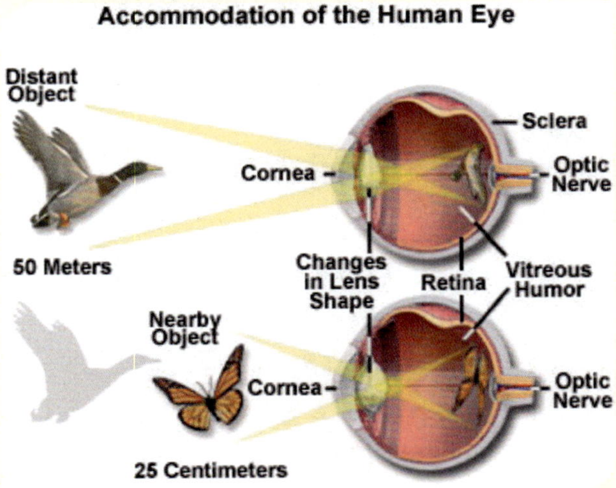

Once light passes through the pupil, it gets focused by a flexible structure called the **lens**. Much like the lens of a camera, its shape can be adjusted to produce images that are clear and in focus. While someone operating a camera adjusts the lens with their hands or an automatic focus, the eyes always use an automatic focus that enlists muscles in the eye to bend the lens and bring images into focus. It's interesting to note that due to the way the lens bends the light rays coming in, the images produced are upside-down and reversed. (This is the same way images enter and are recorded by cameras.) The brain is then able to record this awkward image when the light strikes the layer of receptor cells that line the back of the eye, called the **retina** (ret-n-uh).

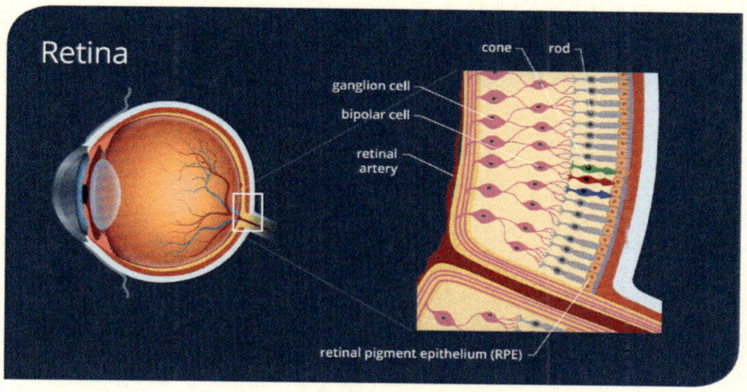

The retina of each eye has 130 million cells, designed to detect light as it moves into the eye, called photoreceptor cells. They send signals to the brain about what the eyes are looking at, via the optic nerve (more about that soon). Within the brain, the signals carried by the optic nerve are translated (processed) into the images we see of the world around us.

The vast variety of colors and sights that we see are formed by two different types of the light-gathering and recording photoreceptor cells: rods and cones. **Rods** do not distinguish between different colors, but they are extremely sensitive to light. They allow people to see objects when it is dark, but only in shades of gray. **Cones**, on the other hand, are capable of recognizing and responding to three different colors: blue, red, and yellow. By blending these three colors, your cones enable you to see in color.

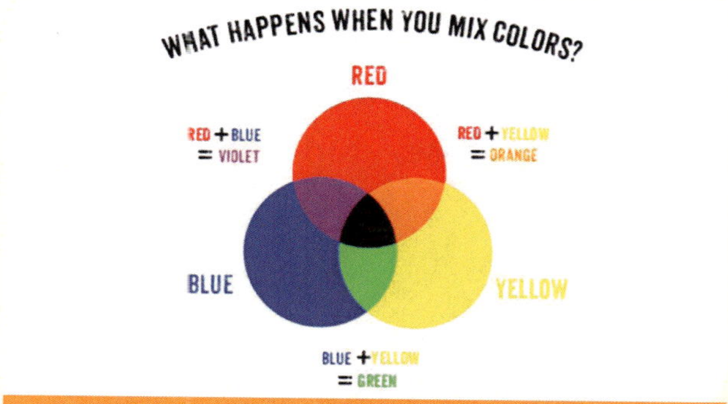

Any color can be made by mixing the three primary colors that the eye sees.

If you are wondering how you can see so many different colors when your eyes can only recognize blue, red, and yellow, take a look at the ink cartridge of a color printer. It has only three colors, and yet it can print virtually any picture. That's because different colors can be blended together to make other shades and colors. Similarly, the brain can "put the colors in place" and recognize the beautiful sights all around you.

Understanding the two different types of photoreceptors helps us understand the condition of color blindness. People who are color-blind do not have fully functional cones, but their rods do work. That's why they can see objects but cannot make out their colors.

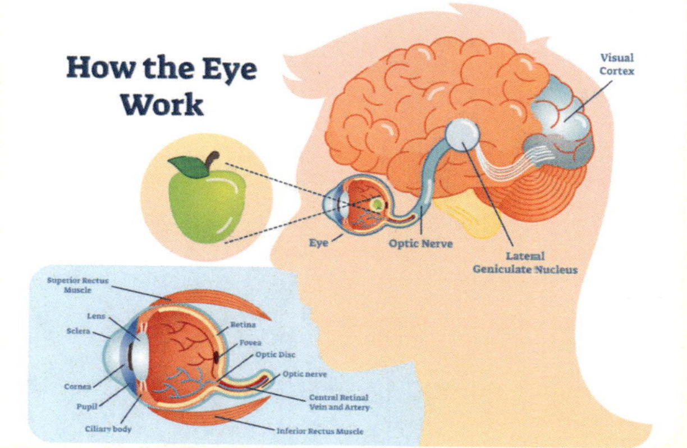

All the impulses generated by rods and cones leave each eye by way of an optic nerve that passes through the back of the eye. Each optic nerve, one for each eye, then carries the impulses to the section of the brain that was designated to interpret them and translate them into visual images so you can enjoy the gift of sight.

Why People Wear Glasses

Are you one of the many people who has trouble seeing without the aid of eyeglasses?

Approximately 20% of children and 64% of adults in the United States wear glasses. Let's learn a bit about why people need glasses and how glasses work.

Lenses, both those found in your eyes and the ones found in eyeglasses, are made out of curved transparent materials that bend light as it passes through them. Vision problems arise when the lenses of the eye do not focus light directly onto the retina. This occurs when the eyeball is a tiny bit longer or shorter than usual. In such situations, the light must travel a longer- or shorter-than-usual distance before reaching the retina, which causes the image to be distorted.

When the eyeball is longer, light must travel a longer distance before reaching the retina—making distant objects appear blurry. This condition is called **nearsightedness** because people with it can see nearby objects clearly, but not those that are far away. **Farsightedness** occurs when the eyeball is shorter; light from nearby objects gets bent and distorted, while objects far away can be seen clearly.

In both cases, commercial lenses can be used to correct the problem and enable people to see clearly. Depending on whether the person is nearsighted or farsighted, the shape of the lens can adjust the light so it is focused properly on the retina and a clear image is sent to the brain.

In the case of nearsightedness, the shape of the eye causes images to project in front of the retina but not directly on it, as is the case with people with perfect vision. To correct this, they wear concave lenses (those that arc outward), as they bend the light away from the center of the lens. This spreads out the light before it reaches the lens of the eye, causing the resulting image to project a bit further back and land directly on the retina, thus providing clear vision.

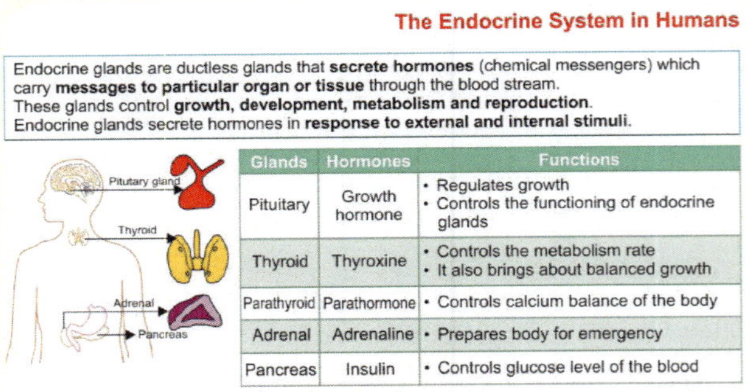
The Endocrine System in Humans

Endocrine glands are ductless glands that **secrete hormones** (chemical messengers) which carry **messages to particular organ or tissue** through the blood stream.
These glands control **growth, development, metabolism and reproduction**.
Endocrine glands secrete hormones in **response to external and internal stimuli**.

Glands	Hormones	Functions
Pituitary	Growth hormone	• Regulates growth • Controls the functioning of endocrine glands
Thyroid	Thyroxine	• Controls the metabolism rate • It also brings about balanced growth
Parathyroid	Parathormone	• Controls calcium balance of the body
Adrenal	Adrenaline	• Prepares body for emergency
Pancreas	Insulin	• Controls glucose level of the blood

How glasses work

In the case of farsightedness, the shape of the eye causes the image to project behind the retina, and a convex lens is used to correct this. Convex lenses curve inward, which makes the light rays bend toward each other before reaching the lens of the eye. This results in the image appearing closer than it otherwise would, landing directly on the retina, producing a clear image.

Hearing and Balance

While everyone knows that the ear is the organ that enables you to hear, the fact that it also helps you keep your balance is often less understood. First, we will learn more about hearing and then we'll explain the ears' role in balance. The ear is made up of three different sections that work together to collect sounds, convert the sounds to electrical impulses, and send the impulses to the brain for processing.

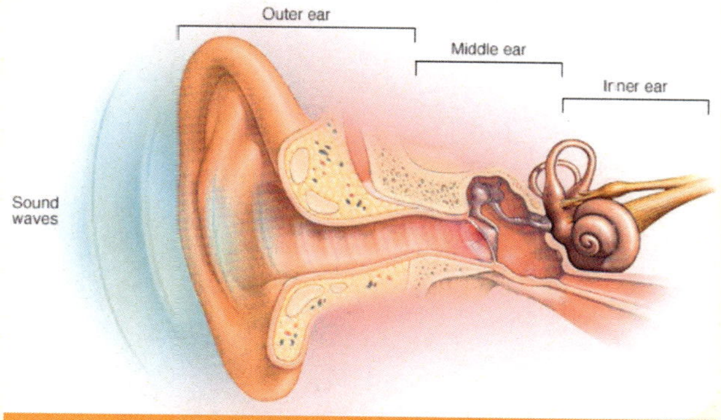

Sections of the ear

The three parts of the ear are the outer ear, the middle ear, and the inner ear. The outer ear is the part of the ear that you see. It has a funnel-shaped design so that it can better fulfill its function, which is to gather sounds and funnel them to the ear canal—the pathway leading to the middle ear. The ear canal also contains glands that secrete earwax. That seemingly useless wax actually protects the ear by preventing dirt from entering (think of a sticky fly trap) and by helping to prevent ear infections. The outer ear is not made of bone; rather, it is tough, skin-covered cartilage. That's why it is bendable.

Next comes the middle ear, which is an air-filled cavity that turns sounds into vibrations and delivers them to the inner ear. The **eardrum** is made up of a membrane (thin flap of skin) that is stretched tight. It works very much like a drum, which vibrates when it is struck. When sound "strikes" the eardrum, its membrane vibrates, and those vibrations are passed on to three tiny bones—the smallest bones in the body—called the hammer, anvil, and stirrup. These bones are all named for their shapes. When sound first enters the ear, it makes the eardrum vibrate, then the vibrating eardrum makes the hammer vibrate, which then makes the anvil vibrate, which, last but not least, makes the stirrup vibrate.

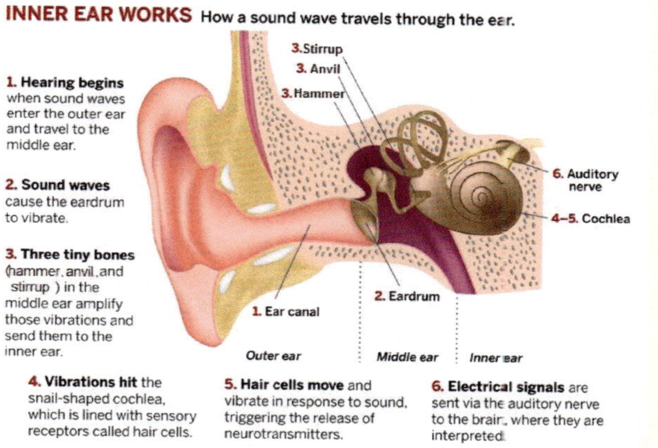

Beyond the stirrup lies the inner ear, where the vibrations passed along the middle ear are turned into nerve signals. The inner ear includes the cochlea (koh-klee-uh) and the semicircular canals. The **cochlea** looks a lot like a snail. It changes the vibrations from the middle ear into nerve signals by way of tiny hairs that are moved by the sound vibrations and that trigger the nerves to fire. These signals travel along the auditory (hearing) nerve to the brain, where the signals are translated into the sounds you hear.

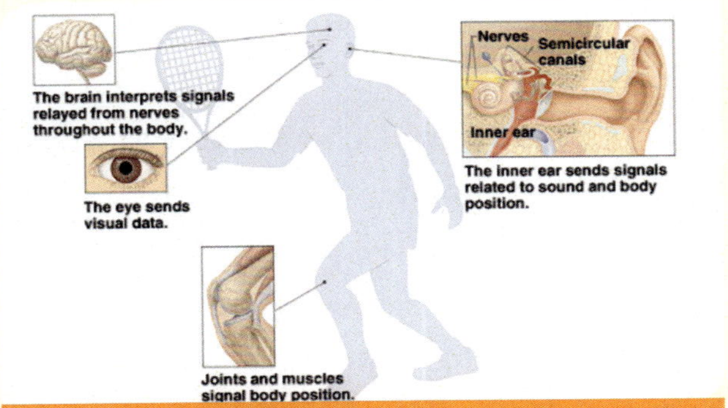

Staying balanced

The **semicircular canals** look like three tiny, connected tubes. It is their job to help you keep your balance. Staying balanced might seem pretty simple, but if you spin yourself around a few times you'll notice that there's more than meets the eye to staying straight. The system Hashem designed to keep you balanced consists of these three small semicircular canals, which are filled with fluid and lined with tiny hairs. When your head moves, the fluid in the canals swishes around, moving the hairs. The hairs' movement triggers nerves to send information on your position—such as if you are standing up straight or lying on your side—in the form of signals to the cerebellum, the part of the brain in charge of keeping you balanced. The cerebellum then interprets these signals and sends messages to the muscles in your limbs about how to position your body in order to stay balanced.

The science behind the spin

When you spin around and stop, the reason you feel dizzy and have trouble walking straight is because the fluid in your semicircular canals continues to swish around for a while, giving your brain the idea that you're still spinning even when you aren't. When the fluid stops moving, the dizziness goes away.

Healthy Eyes and Ears

Your eyes and ears are extremely important and have delicate structures that must be taken care of if you want to keep these organs working their best.

For the health of your eyes, one of the most important things is to get sufficient vitamins and minerals that strengthen eyesight. Vitamins A, C, E; omega-3 fatty acids; zinc; and beta-carotene are all beneficial to your eyes. Examples of foods where these nutrients are found include: cabbage, spinach, and kale, rich in vitamin A; oranges and lemons, rich in vitamin C; almonds, peanuts, and sunflower seeds, rich in vitamin E; salmon and tuna, rich in omega-3 fatty acids; eggs, grapes, and dark berries, a good source of zinc; and carrots, packed with beta-carotene.

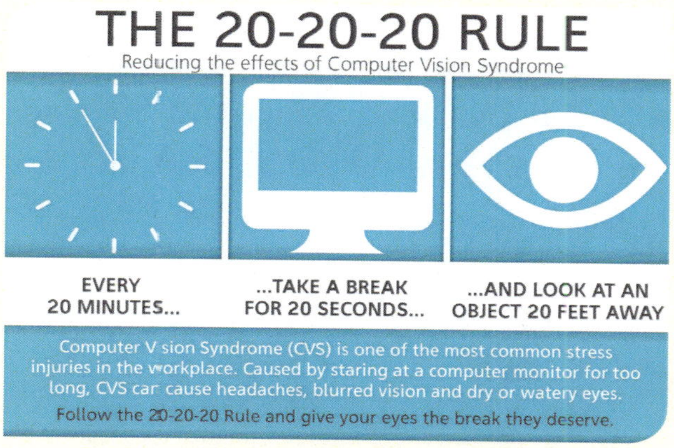

Another way of helping your eyes stay healthy is to make sure that the light is not too bright or dim when you are working. Lower-intensity bulbs and fluorescent lights are an ideal way to lighten your work/study area. In addition, if you are working on a computer, make sure that its brightness, text size, and the contrast (proportion of brightest and darkest colors) properly suit your vision and make you feel comfortable. That's important, since many people who work at a computer for long stretches of time suffer from CVS, computer vision syndrome. CVS can cause headaches, blurry vision, and watery eyes. Eye doctors advise those using computers to look up and away from the screen once every 20 minutes to relax the muscles inside the eyes.

The Problem with Cotton Swabs

To keep your ears healthy, many doctors advise against sticking things like cotton swabs and fingernails deeply into the ears. Attempting to use them to clean out ear wax can scratch the ear canal, push earwax deeper into the ear, and even rupture the eardrum. If earwax bothers you, it is best to seek the advice of a doctor to help remove it safely.

Another way to protect your ears is to avoid loud noises. Loud noises can cause permanent hearing damage, so be careful. Listening to music at high volumes on portable music players, at concerts, at *simchah*s, or anywhere else, can do a lot of damage. Tiny earbuds can actually harm your hearing as much as standing near a large, noisy chainsaw if the volume is turned up to the maximum. If you must be in the presence of loud noises, it's advised to wear earplugs to minimize the effects.

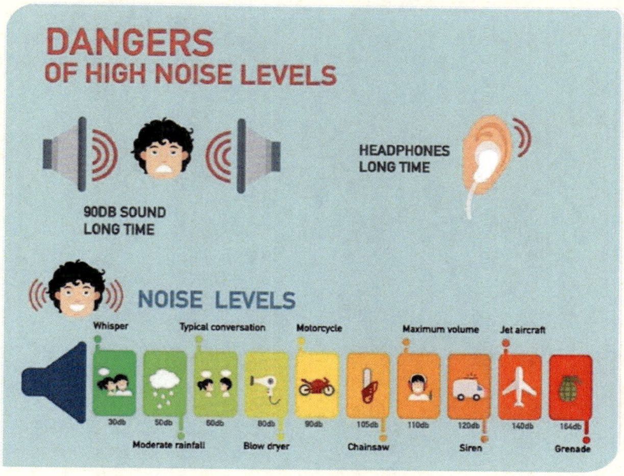

Some people don't think that their hearing is being damaged by noise because they were once at a very loud music event and they can still hear just fine. In truth, damage did occur. The reason why they don't immediately feel the effects is that hearing damage builds up over time. So, while the occasional wedding or other *simchah* where loud music is being played will not do too much damage, you must be careful to protect your ears and avoid regular doses of high decibels (units of sound measurement)

Smell and Taste

The senses of smell and taste have much in common and often work closely together. Both smell and taste respond to chemicals. Receptors in the nose respond to chemicals in the air, and receptors in the mouth respond to chemicals found in foods. In both cases, when chemicals land on receptors, nerve impulses are generated and sent to the brain. That's where they are translated into the many different tastes and smells you are familiar with.

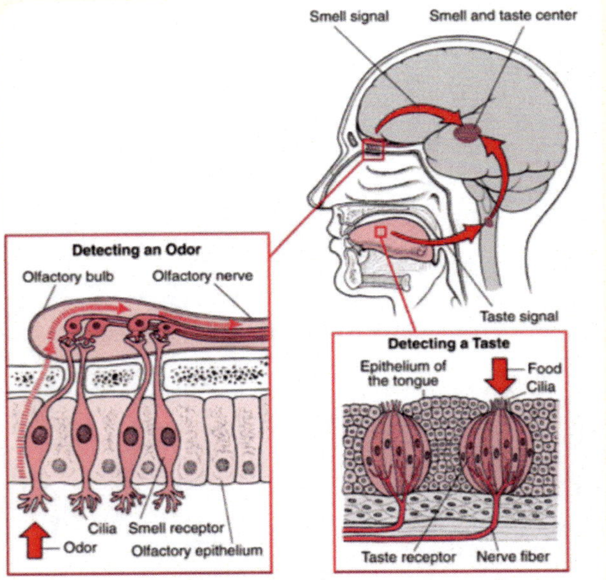

The senses of taste and smell

The nose can distinguish at least ten different categories of odors, ranging from peppermint to pungent, which include at least 10,000 different distinctly identifiable scents. Recent studies have claimed that the number of scents people are able to identify is much higher than the 10,000 previously identified. These studies estimate the number is over a trillion, but others challenge that claim and the mathematics used to arrive at the giant number.

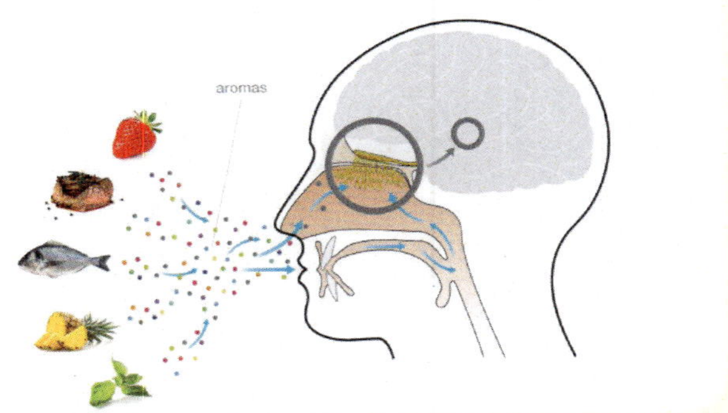

Whatever the actual number is, it is certainly a lot and many more than the number of tastes that can be recognized by the tongue. There are five basic tastes the tongue can detect: sweet, sour, salty, bitter, and umami (a meat-like taste). If you take a good look at your tongue, you'll notice there are many little bumps on it. Those bumps are called taste buds. Cells capable of detecting tastes are located on the sides of the taste buds, and different areas of the tongue have higher concentrations of taste buds than other areas. While all taste buds can detect the five tastes, different parts of the tongue detect certain tastes better than others.

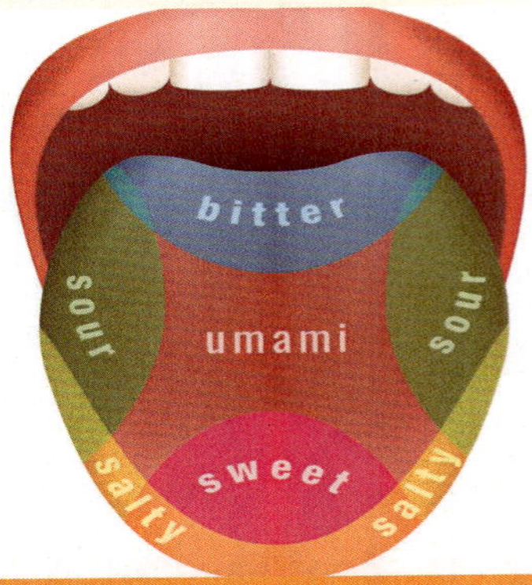

Areas of taste on the tongue

Even though there are only five basic kinds of tastes that can be detected by the tongue, there is a wide variety of tastes that can be experienced, because the flavor of foods is influenced by both its smell and its taste. This fact is most noticeable when someone has a cold and a stuffy nose. That's when people often find that foods have "lost their taste." But the real reason is because they can't smell the food, and smell contributes to our experience of taste.

Touch

You learned a lot about the sense of touch in chapter 4, in the section on skin, but there are some more facts that you will find interesting. The sense of touch is the largest of the sense organs because it is not located in any one area, but all throughout the body wherever there is skin. There are many kinds of touch receptors, which respond to different stimuli such as pressure, temperature, and pain.

Touch receptors are located at different depths. The ones closer to the surface of the skin can detect light touch, such as when something brushes against you or when you feel the texture of objects ranging from rough to smooth. The receptors located further down in the dermis detect feelings of pressure. These receptors are not triggered by light touch. However, when we exert force—such as when we start lifting a heavy object—they are fired, alerting the brain as to about how heavy an object might be, and if lifting it is a good idea.

The sense of touch, along with those of sight, hearing, smell, and taste, are the windows Hashem gave us to experience the world. These gifts are so incredible and precious, it's very hard to ever thank Him enough for them. To help us appreciate this, we'll conclude this section with a very relevant paragraph written by Rabbi Avigdor Miller, in *Ohr Avigdor*, *Chovos HaLevavos*, *Shaar HaBechinah* (pp. 243–244):

Although every detail of the body is a *chesed* that is designed to help us, the most important function of the body is that we should look at it and marvel at the *chesed* of Hashem. We therefore have to study our bodies as much as possible and sing to Hashem out of gratitude that He has given us eyes. And in order to spur us on to greater enthusiasm, from time to time He gives us an opportunity to see a man on the street who has no eyes. When we see a blind man tapping his way with a cane or when we see people with bandages on their eyes, we shouldn't let that opportunity go to waste. We should think how fortunate we are that Hashem gave us eyes.

SECTION 5.2 REVIEW

1. What are the five senses?

____

2. What are the functions of the cornea, iris, pupil, and lens?

____

3. What are rods and cones, and in what ways are they different?

____

4. How does the ear take in sounds? Which structures do the sounds pass through before they are turned into nerve impulses that are sent to the brain?

____

5. Since there are only five different taste receptors on the tongue, how can we explain the wide variety of different tastes found in foods?

____

USE YOUR MIND

What is the most important sense to you, and why?

____

____

SECTION 5.3: HORMONES AND THE ENDOCRINE SYSTEM

See if you can guess what is causing the following scenario. Someone's heart is racing, their hands are sweating, and they feel the blood pumping through their veins. While this reaction could result from many different situations, such as when one is forced to fight off a foe, about to speak in public, or involved in an intense sporting event, the underlying cause of these bodily sensations is due to a chemical that the body releases, called adrenaline. Adrenaline is one of many chemical messengers called **hormones** that are released in order to regulate bodily activities.

Members of the police force and soldiers engaged in combat feel the rush of adrenaline before engaging their enemy.

Hormonal Activity

You might be wondering how hormones affect only some organs and not others. How do they hit their targets? The answer is found in their chemical structure, which interacts only with cells specific to that particular hormone. Those cells are called **target cells** and they recognize the hormone's structure. Like a key fitting into a lock, hormones travel through the bloodstream until they find the "lock" (particular target cell) they were designed to react with.

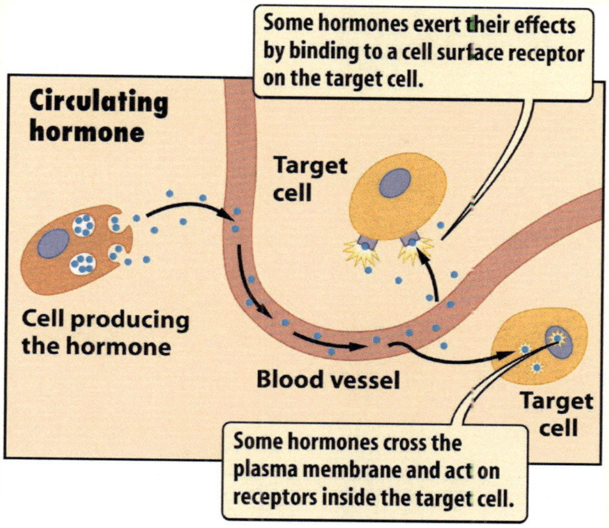

Although nerve impulses create an almost immediate reaction, the release of hormones takes more time to elicit a response because they first have to travel through the blood to their target organs. Similarly, when the hormones are no longer being released, their effect will continue to be felt for a longer time, until the level of hormones in the bloodstream has dropped significantly. It is for this reason that it often takes a person a little while to "calm down" after they've gotten a rush of adrenaline from being in a fight or other intensely stressful situation.

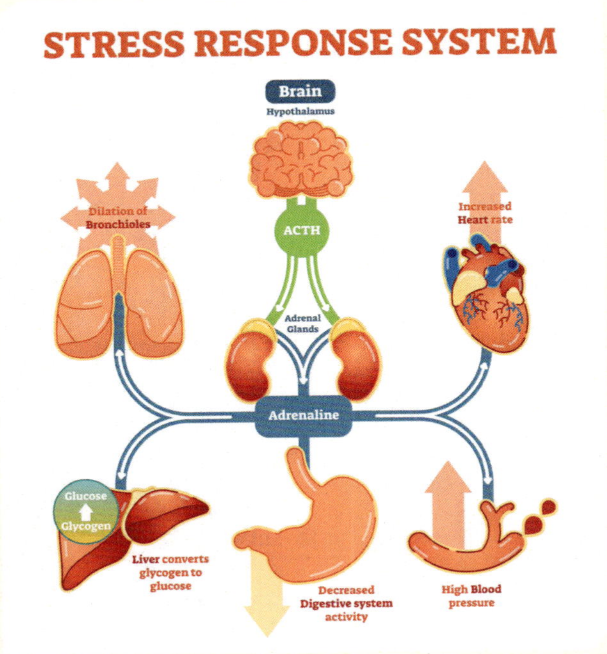

Adrenaline is the hormone responsible for the body's response in situations where survival is threatened (or it feels that way to us) and crucial decisions must be made on the spot. This is the fight-or-flight reaction mentioned earlier. This reaction causes the air passages of the lungs to become larger, providing the muscles with more oxygen to either fight back or run away (or, in the case of public speaking, to either give that speech or run off stage!). Adrenaline triggers the blood vessels to contract so that blood can be transported to major organs in the body, like the heart and lungs. It also increases a person's ability to tolerate pain, which is why an individual or even an animal can often still run or fight, even after sustaining injuries. Adrenaline also causes a noticeable increase in strength, speed, and performance of virtually any physical activity, as well as enhanced awareness, especially during stressful times.

Endocrine Glands

Hormones are produced by a number of **endocrine** (en-duh-krin) **glands** located at different places in the body. These glands release their chemical messengers into the bloodstream. Once in the bloodstream, hormones can travel quickly to areas of the body far from the glands they came from, and affect the activities of organs and tissues in a number of ways, such as turning them on or speeding up their activity. In fact, these chemical messengers control most major bodily functions, from simple basic needs like hunger to more complex systems like emotions and moods. All the endocrine glands in the body, along with the hormones they produce, are collectively referred to as the **endocrine system**. This system controls many of the body's daily activities in addition to its role in regulating many long-term changes such as growth and development.

The Endocrine System in Humans

Endocrine glands are ductless glands that **secrete hormones** (chemical messengers) which carry **messages to particular organ or tissue** through the blood stream.
These glands control **growth, development, metabolism and reproduction**.
Endocrine glands secrete hormones in **response to external and internal stimuli.**

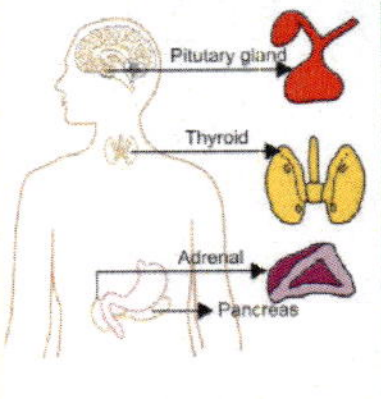

Glands	Hormones	Functions
Pituitary	Growth hormone	• Regulates growth • Controls the functioning of endocrine glands
Thyroid	Thyroxine	• Controls the metabolism rate • It also brings about balanced growth
Parathyroid	Parathormone	• Controls calcium balance of the body
Adrenal	Adrenaline	• Prepares body for emergency
Pancreas	Insulin	• Controls glucose level of the blood

Some of the endocrine glands and their functions

The endocrine system works together with the nervous system to regulate the body's activities—with the nervous system using electrical signals to communicate and the endocrine system using hormones. The two systems usually work together to release hormones. For example, after seeing that a potential danger is approaching, the brain will send nerve impulses to an endocrine gland, which gets the signal to release adrenaline into the bloodstream. Adrenaline then makes the heart rate increase and breathing speed up, to get the person ready for a fight-or-flight situation.

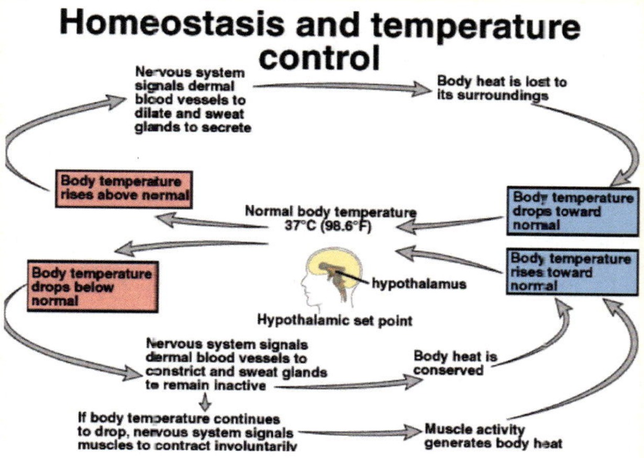

How the hypothalamus works to maintain homeostasis

One of the glands, the **hypothalamus** (hay-puh-thal-uh-muhs), plays a primary role in maintaining your body's internal balance (homeostasis). The hypothalamus is found in the lower central part of the brain and receives many nerve impulses carrying messages about the surrounding temperature, light exposure, and feelings of the body. It responds to that information by stimulating or inhibiting many of your body's main processes such as heart rate, body temperature, and hydration (water) level.

The hypothalamus is also considered the link between the nervous system and the endocrine system, primarily due to its control over the **pituitary** (pi-too-i-ter-ee) **gland**, often considered the "master gland." The pituitary gland is located at the base of the brain. In response to cues from the hypothalamus, it sends hormones that activate other glands to release their hormones. Some of the hormones the pituitary gland makes are: growth hormone, which stimulates the growth of bone and other body tissues and plays a role in the body's handling of nutrients and minerals; antidiuretic (an-ty-dy-uh-reh-tik) hormone, which directs the kidneys to regulate the amount of water in the blood; and endorphins (en-dor-fins), which are hormones that are nicknamed the "happy hormones" because they act on the nervous system to produce feelings of pleasure and reduce feelings of pain.

Negative Feedback

Seeing that the release of hormones is necessary to maintain homeostasis, and they can affect a wide variety of reactions in the body, an extremely efficient system was designed to regulate their release. The type of system Hashem designed to regulate hormone release by the endocrine system is called negative feedback or a negative feedback loop. **Negative feedback** is called negative because the system is turned off by the condition it produces.

The most common example given of negative feedback is the thermostat in an air-conditioning system. To explain how the system works, let's say it's a hot summer day and you come home to cool off. You set the air conditioner to 72 degrees Fahrenheit (22 degrees Celsius). When the AC system senses that the temperature is above 72, it blows cool air into the room. When the room reaches the desired temperature, this information is "fed back" to the thermostat, which then sends a signal to the cooling unit to turn off.

Negative Feedback (Example)

1. You set the temperature (let's say 72 degrees)
2. It get's too hot in the room.
3. The Air conditioner turns on.
4. The temperature lowers to 72.
5. The air conditioner turns back off.

In a similar way, the endocrine system uses negative feedback to maintain proper levels of hormones and maintain homeostasis. When the amount of a particular hormone in the blood reaches the desired level, the endocrine system sends signals that stop the release of that hormone. A great example of this is production of the hormone insulin, which happens every time you eat.

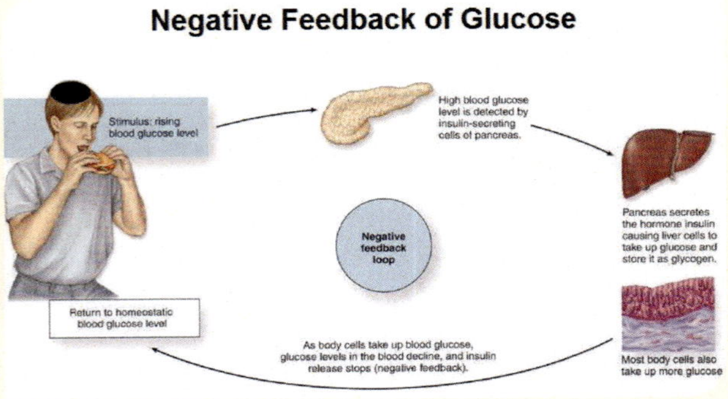

Insulin is produced by the pancreas. The pancreas causes cells to take in more glucose, and causes the liver to convert glucose into a storable form called glycogen, which is made up of many glucose molecules. When food enters the body after a meal, glucose levels rise. This triggers the pancreas to release insulin, which causes the levels of glucose in the body to go back down. When the glucose reaches the desired level, insulin production is turned off, just like the AC shuts off when the temperature reaches the right level. Regulatory mechanisms like that which maintains your blood glucose levels are at work all the time, keeping the miracle machine of your body working well day and night.

With this, we conclude our tour of the human body—but really, our tour will continue until 120, as we travel through life in an amazing, customized vehicle that Hashem designed to give each one of us the best possible ride!

SECTION 5.3 REVIEW

1. What are hormones?

____

2. What is the endocrine system made up of, and what does it do?

____

3. What is the hypothalamus, and what does it do?

____

4. Why is the pituitary gland nicknamed the "master gland"?

____

5. What is negative feedback? Give one example of it.

____

USE YOUR MIND

Name an advantage and a disadvantage of the endocrine system, compared to the nervous system, as a means of communication for the body.

____

____

SECTION 5.4: EXPAND YOUR MIND

SEEING THE PLAN AND THE PURPOSE

The human brain is one of, if not the, greatest feats of creation in the universe. Just how spectacular the brain is and how much information it is capable of processing is overlooked and underappreciated by many. To help us better appreciate Hashem's masterpiece of design and engineering, let's try comparing it to another information storage and processing unit we're familiar with that lies outside of our bodies—a computer.

According to an article written by Dr. Paul Reber, professor of psychology at Northwestern University:

The human brain consists of about one billion neurons. Each neuron forms about 1,000 connections to other neurons, amounting to more than a trillion connections. If each neuron could only help store a single memory, running out of space would be an issue. You might have only a few gigabytes of storage space, similar to the space on a USB flash drive. But, neurons team up so that each one helps store many memories at a time, exponentially increasing the brain's memory storage capacity to something closer to around 2.5 petabytes (or a million gigabytes).[1]

In comparison, the average computer can only store about 500 gigabytes of information. That means your brain can store the same amount of information as two thousand computers combined! You might be wondering: If the brain can store so much information, why am I worried about what grade I'll get on my upcoming test on the nervous system? There's enough room to store this entire textbook in our minds, with tons of room to spare.

Although it is true that the brain records and remembers virtually everything we see or do, those memories are not accessible to our conscious mind. The *sefer Chovos HaLevavos* explains one of the reasons for this is that we would be overcome with grief if we never forgot any of the hard things that ever happened to us.

Now, 2.5 petabytes only takes into account the information that can be stored in one's memory. Dr. Werner Gitt, former head of the Department of Information Technology at the German Federal Institute of Physics and Technology, explains that "if we take all human information processes together, i.e., conscious ones (language, information-controlled, deliberate voluntary movements) and unconscious ones (information-controlled functions of the organs, hormone system), this involves the processing of 10^{24} bits daily. This astronomically high figure is higher by a factor of 1,000,000 [i.e., is a million times greater] than the total human knowledge of 10^{18} bits stored in all the world's libraries."[2]

So, the human brain is clearly the most sophisticated information storage and processing system in the universe, and we have one right in our very own head! Now think about the chemical factories of the liver, stomach, and other places that are more intricate than any factory that man has ever built. They transform food into blood, bones, and muscles! Together with the automatic heating and cooling system of the skin; the automatic-focus cameras of our eyes; the sound-recording and balance mechanisms of the ears; the mighty, muscular pump of the heart, which beats nonstop without ever getting tired; and many other Niflaos HaBoreh we have learned about together, we now have a much easier time saying David HaMelech's words with great *kavanah*: "I shall thank You, for in an awesome, wondrous way I was fashioned; Your works are wondrous, and my soul knows it very well" (*Tehillim* 139:14).

1 Paul Reber, "What Is the Memory Capacity of the Human Brain?" *Scientific American*, May 1, 2010.
2 Werner Gitt, "Information: The Third Fundamental Quantity," *Siemens Review*, 56, no. 6 (Nov.–Dec. 1989).

SECTION 5.4: EXPLORE CHAZAL

TORAS EMES

We learn in the *Shulchan Aruch* (*Orach Chaim* 25:5) what a person must remember when he wears tefillin—that Hashem commanded us to strap these four Torah passages containing the oneness of His Name and the exodus from Egypt on the arm, which corresponds to the heart, and on the head, corresponding to the brain, so we remember the miracles and wonders that Hashem performed for us and His oneness that he displayed to us—demonstrating His supreme power and control over the upper as well as the lower worlds. And man must subject to Hashem his *neshamah* (soul), which resides in his brain, as well as his heart, which is the main area of his thoughts and desires.

The four chambers of the tefillin worn on the head contain the Torah passages corresponding to the four lobes (sections) of the brain. When a person performs this mitzvah, he is placing the tefillin at the physical "gate" of the soul that rests within the brain. This allows the *neshamah* to connect to Hashem's eternal light, which infuses it with illumination. Similarly, the tefillin of the arm, which rest upon the bicep muscle angled toward the heart, transmit their light to the *nefesh* (spirit) that resides within the heart. The reason that the *neshamah* is located in the brain is because it is the source of human thought and logic; the *nefesh* is in the heart because it is the source of all human emotion. They are the spiritual "life support" that sustain our physical heart and brain.

Another interesting aspect of tefillin is brought in an article by Rabbi Judah Dardik. He recalls how his bar mitzvah teacher had suggested that tefillin serve as a sort of "spirituality receiver." He writes:

At this very moment there are radio and cellphone and television transmission waves going through the room, yet no one hears them for lack of a receiver. So, too, [the teacher] suggested, there are "spirituality waves" present around us, but they may require some sort of a "receiver" to bring them to consciousness. He identified tefillin as just this sort of tool that can heighten sensitivity to connection with the infinite.

This seemed to me an interesting idea that could not be proven either way. Then in graduate school I ran across an article on the front page of the New York Times Science Tuesday section. The article had nothing to do with Jewish practices or theology; it was an article about brain mapping.

The narrative portion of the article spoke of the famous case of Phineas Gage. Gage was a foreman who helped to lay track for the Burlington Railroad, and an accidental explosion sent a metal rod up through his jaw and head. To the astonishment of those around him, he not only survived the blast but was reasonably functional afterward. Yet the article noted what interested researchers, namely, that he seemed to lose his moral compass and didn't care about right and wrong anymore. The researchers mentioned were using medical records of cases (including Gage's) as well as PET scans to ascertain the seat of moral thinking in the brain.

What blew me away were the pictures that accompanied the Times article. They showed from several different angles the spot of Gage's brain damage that was under consideration, and it was precisely where Jews have been donning tefillin for thousands of years: centered between the eyes and just above the hairline. There before me lay the connection between a spot in the brain noted for the kinds of thoughts that underlie ethical decision-making and my teacher's concept of a spirituality receiver. I was stunned.[1]

In relation to the "spiritual receiver" ability of tefillin, it's interesting to note that an article by Steven Schram, a chiropractor and acupuncturist (person who practices a certain type of Chinese medicine) mapped out several traditional methods of binding tefillin—and noted major underlying acupuncture points. Schram's investigation determined that when bound, contact points of tefillin are the same pattern of an acupuncture treatment to improve mental clarity and spiritual consciousness, and purify thoughts.[2]

Time after time, we see that the more we learn about science and nature, the more the infinite wisdom of the Torah is revealed.

---

1 Judah Dardik, "Torah: The Bizarre-O-Meter Can Open the Mind to Deeper Spiritual Wisdom," *The Jewish News of Northern California*, Jan. 27, 2012.

2 Steven Schram, "Tefillin: An Ancient Acupuncture Point Prescription for Mental Clarity," *Journal of Chinese Medicine*, no. 70 (Oct. 2002): 5–8.

EXPERIMENT:

MEASURING YOUR NERVES

Corresponds to Section 5.1

Reminders about Scientific Investigation

A hypothesis is a suggested explanation for a group of facts or phenomena. It is tested by experiments that either support or reject the hypothesis. They are generally presented as "If…, then…" statements.

When conducting an experiment, there are two types of variables: independent and dependent. The independent variable stands alone; it isn't changed by the other variables you're trying to measure. For example, someone's age might be an independent variable. Other factors (such as what they eat) won't change a person's age.

A dependent variable depends on other factors. For example, a test score could be a dependent variable because it could change depending on several factors, including how much you studied, how much sleep you got the night before the test, or even how hungry you were when you took it.

Introduction

How do you know when your hand touches something? How do you know if something brushes against your shoulder? As you learned above, neurons sense things in the environment and send those signals to the brain for analysis. In this lab you'll be exploring an interesting fact about neurons—that they can only send one signal at a time. If two objects touch the same neuron, it can still only send one signal to the brain. This means that a situation could arise where your brain would only sense one thing touching you, even when there were two.

Knowing this, do you think you need the same number of nerves in every part of your body?

Think about your fingertips, arms, and neck. In which of these places might your nerves be closer together—thereby being more sensitive? Make your predictions and see for yourself in this experiment.

Materials

- paper clip
- ruler with centimeters
- blank piece of paper
- pencil
- Hypothesis

Explain which part of your body (fingertip, arm, or neck) will be most sensitive (having the greatest number of nerves) and explain why:

If____,

then____,

because____.

Procedure

1. Open your paper clip until the ends of the clip are exactly 4 centimeters apart.

2. Touch both ends of the paper clip near the tip of your finger. A gentle touch is all that is required. If you feel both ends, write a "2" in the first square in the chart below. If you only feel one paper clip end, that means both ends of the paper clip are touching the same neuron. If this happens, write a "1" in the appropriate square in the chart.

3. Repeat step 2 on your arm and neck, and record your results in your table.

4. Using your ruler, push the paper clip ends 0.5 centimeter closer together. Repeat steps 2 and 3, bringing the paper clip ends closer each time, until they are touching.

5. Use the table below to record your results.

Distance between paper clip ends (cm)	Fingertip	Arm	Neck
4			
3.5			
3			
2.5			
2			

1.5			
1			
0.5			
Ends touching			

Explanation

Every time you got a 2, it means there was more than one nerve in the area that was felt as a separate impulse going to the brain—and that area is more sensitive. When you got a 1, even though there were two points touching your skin, it means there was only one nerve located in that space—and therefore less sensitive.

Questions

1. Which part of the body was least sensitive (i.e., had the most 1s)?

____

2. Which part of the body was the most sensitive (i.e., had the most 2s)? Why do you think that is?

____

3. What was the independent variable in this experiment: the number of points felt or the distance between the clips?

____

4. What was the dependent variable in this experiment: the number of points felt or the distance between the clips?

____

5. Was your hypothesis supported? Why or why not? Use your data to explain your answer.

____

EXPERIMENT:

SENSING YOUR SENSES

Corresponds to Section 5.2

Materials:

- Five pennies
- Small paper cup or rinsed-out yogurt cup
- A blindfold
- 5 Items to make noise (coins to jingle in a jar, a book to close, hands to clap, paper to crumble, paper to rip, a ball to bounce, or supplies for any other sounds you want to make)
- 5 Small, lidded containers that you can't see through
- Smelly stuff, such as lemons, bananas, orange peel, pine needles, a cotton ball soaked in perfume, chocolate, coffee, dirt, vanilla, garlic, onion, mint, vinegar, rose petals, pencil shavings, or ginger
- Foods to taste, such as different flavored jelly beans, different fruit and vegetable slices, or a variety of yogurts and puddings
- Spoons
- A cup of drinking water for everyone
- A pillowcase
- Scissors
- One long gym sock
- 5 Items to fill the pillowcase (for example, cup, spoon, ball, block, fruit, sponge, rock, a small bell, cotton ball, pine cone, feather, wood letters or numbers)

Sight - Introduction:

This experiment is designed to help you appreciate how having two eyes helps you better judge where objects are.

Procedure:

1. Put a cup in front of your partner. It should be about 2 feet away from them.

2. Ask your partner to close one eye.

3. Hold one of the pennies in the air about 18 inches above the table or desk in front of you. Move it around slowly. 4. Tell your partner that you'll drop the penny whenever he or she says, "Drop it!" (The idea is for your partner — with one eye closed — to judge when the penny is over the cup so the penny will drop into the cup).

5. Give your partner five tries with one eye closed, then five tries with both eyes open.

Question:

1. Which way worked best?

____

Sound - Introduction:

This experiment is designed to help you appreciate how many different types of sounds you can recognize.

Procedure:

1. Blindfold your partner.

2. Make each noise.

3. Ask your partner to guess what the noise is.

Questions:

1. What did your partner guess was the noise?

Noise 1______________ Noise 2__________________ Noise 3__________________

Noise 4______________ Noise 5__________________

2. How many did your partner guess right?

____

Taste - Introduction:

This experiment shows how your sense of smell helps you taste foods. Without your nose, you may not be able to tell the difference between foods — especially if the foods all have the same texture (for instance, jelly beans with various flavors, yogurt and pudding, or slices of different fruits and vegetables.

Procedure:

1. Blindfold your partner.

2. Have blindfolded partner pinch their nose.

3. Offer your partner one flavor of the food at a time.

4. Can he or she identify it? (Your partner might want a drink of water between the different foods.)

____

5. Offer the same food and tell your partner to unpinch their nose. Now does your partner know what it is? ____

Smell - Introduction:

This experiment is designed to help you realize how many smells you can recognize. How many? Try this and see.

Procedure:

1. Poke small holes in the container lids (this can be done by the teacher before the class or by students).

2. Put one "smelly" object in each container.

3. Write a number on the side of each container.

4. Tape a label to the bottom of each container to say what's inside.

5. Blindfold your partner and them smell each container.

6. Write down the number of the container and your friend's guess about what's inside.

Questions:

1. Container 1 guess ____________________ Container 2 guess____________________

Container 3 guess____________________ Container 4 guess____________________

Container 5 guess____________________

2. Are the answers right?

____

3. Which are good smells? Which are not?

____

Touch - Introduction:

This experiment is designed to show how it may be difficult to identify objects based only on your sense of touch. Touch combined with sight, hearing, and smell is much more effective. It's even tougher to identify objects if you don't have your full sense of touch, like if you have a sock or glove on your hand.

Procedure:

1. Put all 5 objects in the pillowcase.

2. Blindfold partner

3. Have partner put on gym sock

4. Have partner reach into the pillowcase and try to identify objects.

5. After finishing, show partner what was truly inside the pillowcase.

6. See how easy it is to identify when all senses are restored.

Question:

1. How many items was your partner able to identify?

____

EXPERIMENT:

ENDROCRINE EXCITEMENT

Corresponds to Section 5.3

Introduction:

As we learned in the chapter, hormones help our body know what to do. They are part of our body's communication system – called the endocrine system. The endocrine system helps messages travel throughout our body so that our body can do all the different things it needs to do. There are four main parts to this endocrine system, glands, hormones, bloodstream and receptors. Glands make and send hormones through the bloodstream to specific receptor sites. These hormone messages tell our body to do things like make more blood cells, digest food, absorb vitamins or even grow.

There has to be one specific receptor site for each hormone message, otherwise your glands might tell your stomach to produce more blood cells or your bones to digest food. That would not work! When a hormone matches up with its specific receptor, then the body knows what it is supposed to do next. It is kind of like receiving a letter in the mail. One of the most fascinating things about hormones is how incredibly specific they are!

In fact, hormones act like real puzzle pieces, or like a lock and key. Sometimes, when you are putting together a puzzle, you can force a puzzle piece in next to another piece when it is not really meant to t there. But, hormones and receptors are not quite like that: they have to match up perfectly, or they will not send a message. Today, we are going to do an activity that will help us learn more about how the hormones and receptor sites together like puzzle pieces. You are going to be either a hormone or a receptor. If you are a receptor, you will stand still and wait for your hormone to come and find you. If you are a hormone, you will search around the classroom until you find the receptor that is a perfect match for you. Are you ready? Let's get started!

Materials:

- Labeled puzzle pieces – one per student

Procedure:

1. Create puzzle pieces by cutting shapes out of cardboard and cut them in half, or use pieces from a jigsaw puzzle and label the back (plain) side of them. Each pair of puzzle pieces should be labeled: with an H on one half for hormone, and with an R for receptor on the other half.

2. Write an action across both pieces, so that it can only be fully read when they are joined together. Possible actions include: jump up and down, jump on one foot, put your hand on your head, turn in a circle, clap your hands, etc..

3. Each student in the class should get one puzzle piece but not to look at the label on their piece.

4. All students should scatter throughout the entire classroom and then freeze after 30 seconds (teacher can count and tell when to freeze).

5. Students then should look at their pieces: students who are "receptors" must remain frozen in place; the "hormone" students may now move throughout the classroom.

6. "Hormone" students must try to match their piece with each receptor piece until they successfully find a match.

7. Once all students have found their match, each pair should act out together the action written on their puzzle pieces.

8. You can repeat the activity if time permits.

Questions:

1. Which body system are hormones are part of?

____

2. What are some examples of body functions that are triggered by hormones?

____

3. Why is it so important that hormones are so specific?

____

GLOSSARY OF KEYWORDS:

active immunity: Protection from a disease that occurs when a person's own immune system produces antibodies in response to the presence of a pathogen

allergy (pl. allergies): An abnormal reaction of the body to an otherwise harmless substance that entered the body, often manifested by itchy eyes, runny nose, wheezing, skin rash, or diarrhea

alveolus (pl. alveoli): One of many tiny air-filled sacs arranged in clusters in the lungs, in which the exchange of oxygen and carbon dioxide takes place

anatomy: The study of the physical structure of animals, plants, or any of their parts

antibody (pl. antibodies): Any of numerous protein molecules produced by B-cells as a primary immune defense. Each molecule has a unique binding site that can combine with the complementary site of a foreign antigen, as on a virus or bacterium, thereby disabling the antigen and signaling other immune defenses.

antigen (pl. antigens): A molecule found on the surface of cells and other substances that the immune system recognizes as either part of the body or a foreign substance

aorta: The main artery coming out of the heart. It branches out to smaller arteries and brings oxygen-rich blood to all parts of the body except the lungs.

arteriole (pl. arterioles): Any of the smaller arteries that connect capillaries to larger arteries

arteriosclerosis: A chronic disease characterized by abnormal thickening and hardening of th arterial walls, resulting in a loss of elasticity and increase in blood pressure

artery (pl. arteries): A blood vessel that carries blood away from the heart

atrium (pl. atria): Either of the two upper chambers on each side of the heart that receive blood from the veins, and in turn force it into the ventricles

autonomic nervous system: The section of the nervous system that controls the involuntary actions of the smooth muscles, heart, and glands

axon (pl. axons): The long threadlike extension of a nerve cell that conducts nerve impulses away from the cell body

bile: A substance produced in the liver and stored in the gallbladder that breaks apart fat particles

blood pressure: Pressure that is exerted by the blood upon the walls of the blood vessels, measured in millimeters of mercury

bolus: A soft, roundish mass or lump, especially of chewed food

brain stem: The part of the brain located at

the base of the brain, which controls and regulates vital body functions including respiration, heart rate, and blood pressure

bronchitis: Inflammation of the bronchial tubes, characterized by coughing and difficulty in breathing, caused by infection or irritation of the respiratory tract by smoking

bronchus (pl. bronchi): Either of the two main branches of the trachea, leading directly to the lungs

calorie (pl. calories): A unit of heat used to indicate the amount of energy that foods will produce in the human body

capillary (pl. capillaries): One of the tiny blood vessels that is one-cell thick, located in between the end of arteries and the beginning of veins. It is the site where substances are exchanged between blood and cells.

carbohydrate (pl. carbohydrates): Any of various substances found in certain foods (such as bread, rice, and potatoes) that provide your body with heat and energy and are made of carbon, hydrogen, and oxygen

carbon monoxide: A colorless, odorless gas formed when a compound containing carbon burns insufficiently. It is present in the exhaust gases of automobile engines and is produced from cigarette smoke and is very poisonous.

cardiac muscle: A specialized form of striated muscle that forms the heart

cartilage: A type of connective tissue that protects the ends of bones due to its flexibility. It is the principal component of structure such as the nose and ear.

central nervous system: The part of the nervous system comprising the brain and spinal cord

cerebellum: A large portion of the brain that coordinates voluntary movements, posture, and balance

cerebrum: The largest part of the brain, consisting of two halves and serving to control voluntary movements and coordinate mental actions

chemical digestion: The breaking down of food particles through chemical reactions

cholesterol: A type of fat. Adequate amounts of it are naturally produced by the body, and if eaten in excess amounts, it can cause clogged arteries and heart disease.

circulatory system: The system that circulates blood through the body, consisting of the heart and blood vessels

cochlea: The spiral-shaped tube, shaped like a snail's shell, that forms part of the internal ear, converting sound vibrations into nerve impulses

compact bone: Hard and dense bone tissue that lies beneath the outer membrane of a bone and above spongy bone

cone (pl. cones): One of many cone-shaped cells in the retina of the eye that are sensitive to bright light and produce color vision

cornea: The see-through membrane of the outer layer of the eyeball that covers the iris and the pupil

cuticle (pl. cuticles): The strip of hardened skin at the base of a fingernail or toenail

dendrite (pl. dendrites): Any of the short, branched threadlike extensions of a nerve cell, which conduct impulses toward the cell body

dermis: The inner or middle layer of the skin, located between the epidermis and hypodermis

diaphragm: The large muscle that separates the lungs from the stomach area and is the principal muscle of respiration

digestion: The process by which food is broken up physically and chemically, and converted into a simpler form

eardrum: The membrane that divides the outer ear from the middle ear

emphysema: A chronic, irreversible disease of the lungs characterized by abnormal enlargement of the lungs

endocrine gland (pl. endocrine glands): A gland of the body that produces hormones and secretes them directly into the bloodstream

endocrine system: The body system that consists of the endocrine glands and the hormones they secrete

enzyme (pl. enzymes): Any of numerous proteins originating in living cells and capable of carrying out chemical reactions in living things

epiglottis: A thin structure that seals the larynx during swallowing, preventing the entrance of food and drink into it

esophagus: A muscular passage connecting the mouth to the stomach

excretory system: The system that sends out wastes from the body

exocrine gland (pl. exocrine glands): Any gland that releases substances outside the body

farsightedness: The condition in which a person sees objects at a distance more clearly than those close by

fat: Energy-containing nutrients that are stored in the soft flesh on people's bodies, helping keep the body warm and protect from injury

femur: A bone in the leg extending from the pelvis to the knee

fermentation: The release of energy from food without using oxygen

food pyramid: A diagram that represents a healthy diet by placing food groups in a pyramid according to the number of servings from each group to be eaten every day

gallbladder: The small organ that stores bile after it is produced in the liver

glucose: A type of sugar that is the major source of energy for cells of the body

hair follicle: A small cavity in the skin from which a hair develops

heart: A hollow, muscular organ that pumps blood throughout the body

hemoglobin: An iron-containing protein found in red blood cells that gives them their red color and transfers oxygen from the lungs to the tissues in the body

histamine: A chemical released by the body in an allergic reaction, causing symptoms of discomfort associated with allergies

homeostasis: The state of stable internal conditions

hormone (pl. hormones): A chemical substance produced in an endocrine gland and transported in the blood to a certain part of the body, on which it exerts a specific effect

hypodermis: The lowermost layer of skin made up mostly of fat tissue

hypothalamus: The part of the brain that controls hunger, thirst, and body temperature. It also regulates various activities of the body and controls the action of the pituitary gland.

immune response: The reaction of an organism's body to foreign materials (antigens), including the production of antibodies

immunity: The ability of an organism to resist disease, either through the activities of specialized blood cells or antibodies produced by them

inflammatory response:

integumentary system: The set of organs that forms the external covering of the body and protects it from various types of damage

interneuron (pl. interneurons): A nerve cell found entirely within the central nervous system that acts as a link between sensory neurons and motor neurons

iris: The colored muscular membrane that surrounds and controls the size of the pupil

joint (pl. joints): Where two bones connect to each other

kidney (pl. kidneys): Two bean-shaped organs that remove waste products in urine and regulate fluid balance

lactic acid fermentation: The process that takes place when oxygen is not present to convert glucose sugar into ATP energy and produces lactic acid

large intestine: The last segment of the digestive system where water is absorbed into the blood and the remaining waste material is prepared for elimination from the body

larynx: The part of the respiratory tract between the pharynx and the trachea; also called the voice box

lens: A transparent structure behind the iris of the eye that focuses light entering the eye on the retina

ligament (pl. ligaments): Strong tissues that hold bones together in movable joints

liver: The body's largest organ. It performs over 500 functions and aids in digestion by producing bile.

lung (pl. lungs): One of the two spongy organs in the chest that serve as the organs of gas exchange

lymph: The colorless fluid that the lymphatic system collects and returns to the bloodstream

lymphatic system: A network of vein-like vessels that returns fluid lost from blood vessels to the bloodstream

lymphocyte (pl. lymphocytes): A white blood cell that functions in immunity and consists of two types, T-cells and B-cells

marrow: The soft tissue that fills the inner

spaces in bone. Red marrow produces blood cells and yellow marrow stores fat.

mechanical digestion: The physical breaking down of food into smaller particles so that it can more easily be processed by the digestive system

melanin: A dark brown coloring found in the body, especially in the skin and hair.

mineral (pl. minerals): Nutrients that are needed by the body in small amounts and are not made by living things

motor neuron (pl. motor neurons): A nerve cell that conducts impulses to a muscle, gland, or other effector

mucus: A protective secretion that lubricates the passage of food in order to make it easier to swallow and to protect the lining of the esophagus and stomach from damage in the gut. It also helps prevent bacterial infection in the nose and throat.

muscular system: The system in the body composed of muscle cells and tissues that bring about movement of an organ or body part

nearsightedness: A condition in which someone is unable to see distant objects clearly

negative feedback: A type of regulation system in which the end product of a process in turn reduces the stimulus of that same process

nephron (pl. nephrons): A microscopic urine-secreting tubule that forms the functional unit of the kidneys

nerve impulse: A wave of physical and chemical excitation that moves along a nerve fiber in response to a stimulus

nervous system: The system of cells, tissues, and organs that regulates the body's response to internal and external stimuli. It consists of the brain, spinal cord, and nerves.

neuron (pl. neurons): A specialized impulse-conducting cell that is the functional unit of the nervous system

neurotransmitter (pl. neurotransmitters): Any of the various chemical substances that transmit nerve impulses across a synapse

nicotine: A colorless, poisonous compound occurring naturally in the tobacco plant

nutrient (pl. nutrients): Substances found in the foods you eat that provide the raw materials and energy the body needs to function properly

osteoporosis: A condition in which the bones of the body become weak and break easily

pacemaker: A group of cells or a body part (such as the sinus node of the heart) that serves to establish and maintain a rhythmic activity

pancreas: A triangular organ that lies between the stomach and small intestine that produces digestive enzymes and certain hormones

passive immunity: Protection from a disease in which antibodies are given to a person rather than produced within the person's own body

pathogen (pl. pathogens): A disease-producing agent, especially a virus or bacterium

pelvis: The large bowl-shaped structure at the lower end of a person's trunk

periosteum: A thick membrane covering the surface of bones

peripheral nervous system: The portion of the nervous system located outside the brain and spinal cord

peristalsis: The progressive wave of contraction and relaxation of muscles, where contents are pushed through a system (such as the digestive tract)

perspiration: A salty, watery fluid secreted by the sweat glands of the skin, especially when very warm as a result of strenuous exertion; also known as sweat

phagocyte (pl. phagocytes): A type of white blood cell that ingests foreign particles, bacteria, and cell debris

pharynx: The passage that leads from the cavities of the nose and mouth to the larynx and esophagus

physiology: The study of the functions of living organisms and their parts

pituitary gland: A small, oval gland attached to the base of the brain, the secretions of which control other endocrine glands and influence growth, metabolism, and maturation

plasma: The liquid part of the blood

platelet (pl. platelets): A cell fragment found in the blood that helps form blood clots

pore (pl. pores): An opening through which sweat reaches the surface of the skin

protein: A substance found in foods (such as meat, milk, eggs, and beans) that provides the materials necessary for tissue growth, repair, and plan an important part of chemical reactions that take place within cells

pulmonary circulation: The circulation of blood between the heart and the lungs

pupil (pl. pupils): The expanding and contracting opening in the iris of the eye, through which light passes to the retina

red blood cell (pl. red blood cells): A cell in the blood that picks up oxygen in the lungs and delivers it to cells around the body

reflex (pl. reflexes): An involuntary response to a particular situation involving no conscious decision

respiration: The process of taking in oxygen from the surroundings and giving out carbon dioxide; also known as breathing

response: Any behavior of a living organism that results from an external or internal stimulus

retina: The membrane lining the inner chamber of the eyeball, containing the rods and cones

rib (pl. ribs): Any one of the curved bones of the chest that connect to a person's spine

rod (pl. rods): Cells in the retina of the eye that are sensitive to dim light but not to color

sclera: The firm white membrane that forms the outer covering of the eyeball

sebaceous gland (pl. sebaceous glands): Any of the glands located in the skin that secrete oily matter for lubricating the hair and skin

sebum: The fatty substance secreted by the sebaceous glands that protects and lubricates skin and hair

semicircular canal (pl. semicircular canals): Any of the three looped tubes of the inner ear that help maintain balance

sensory neuron (pl. sensory neurons): A nerve cell that conducts impulses from a sense organ to the central nervous system

skeletal muscle: Muscle attached to bones that produce all the movements of body parts

skeletal system: The framework of the body, consisting of bones and other connective tissues, which protects and supports the body tissues and internal organs

skull: The bony framework of the head, enclosing the brain and supporting the face

small intestine: The part of the digestive system where most chemical digestion of food occurs and nutrients are passed into the bloodstream

smooth muscle: Muscle tissue that contracts without voluntary control

somatic nervous system: The part of the nervous system that controls voluntary movements in the body

spine: The series of vertebrae extending from the base of the skull to the tip of the tail bone

spongy bone: A layer of bone tissue with many small spaces; located just under the layer of compact bone

stimulus (pl. stimuli): Any activity in the outside environment or within the body that can cause a reaction by the body

stomach: A muscular pouch located in the abdomen where the majority of mechanical digestion of food occurs

striated muscle: Muscle tissue in which the contractile fibers in the cells are aligned in parallel bundles, Muscles of this type are under voluntary control.

synapse (pl. synapses): A region where nerve impulses are transmitted and received

systemic circulation: Circulation of blood from the heart to the rest of the body (except the lungs) and back

tar: A solid, sticky substance that remains when tobacco is burned

target cell (pl. target cells): A cell selectively affected by a particular virus, drug, or hormone

tendon: A band of tough tissue that connects muscle to bone

trachea: The tube extending from the larynx to the bronchi, serving as the principal passage for conveying air to and from the lungs; also referred to as the windpipe

ureter (pl. ureters): The tube that carries urine from the kidney to the urinary bladder

urethra: The passage that moves urine from the urinary bladder out of the body

urinary bladder: A saclike structure that stores urine until it can be passed out of the body

urine: The liquid waste matter excreted by the kidneys; a yellowish, slightly acidic, and watery fluid

vaccine (pl. vaccines): A preparation of a weakened or killed pathogen, such as a bacterium or virus, that stimulates antibody production against the pathogen but is incapable of causing severe infection

vein (pl. veins): A blood vessel that carries blood toward the heart

vena cava: Either one of the two large veins that convey oxygen-poor blood to the heart

ventricle (pl. ventricles): Either of the two lower chambers on each side of the heart, which receive blood from the atria and in turn force it into the arteries for distribution around the body

venule (pl. venules): Any of the smaller veins that connect capillaries to larger veins

vertebra (pl. vertebrae): The small bones that make up the spine

villus (pl. villi): One of many tiny finger-shaped structures that cover the inner surface of the small intestine and provide a large surface area through which digested food is absorbed

virus (pl. viruses): An extremely small particle that causes a disease and spreads from one person or animal to another

vitamin (pl. vitamins): Molecules that are found in small quantities within the body and serve as helper molecules for a variety of essential chemical reactions needed for staying healthy

vocal cords: The two folded pairs of membranes in the larynx that vibrate when exhaled air that passes through them, producing sound

white blood cell (pl. white blood cells): A colorless or white cell in the blood that helps the body fight infection and disease

Made in United States
Orlando, FL
04 November 2026

90205906R00121